AF540823

MODERN DICTIONARY

PHYSICAL CHEMISTRY

MODERN DICTIONARY

PHYSICAL CHEMISTRY

M. Satake
T. Nagahiro

1990

Discovery Publishing House
New Delhi

First Published 1990

ISBN-81-7141-114-2

Published by
Discovery Publishing House
4594/9, Darya Ganj
New Delhi 110 002

Printed in India
at
Taruna Enterprises Composed by Amit Enterprises,
Anup Market, Maujpur, Delhi-110053

PREFACE

With more and more specialization and expansion of the subjects, it has become essential to bring out the dictionaries in these special fields, particularly in modern sciences. Keeping this in view, the present volume has been written to provide students at undergraduate level, a concise view of the subject in the form of a dictionary. The entries have been made in a clear and lucid manner to provide both straight forward definitions and essential details. Figures have been included, wherever necessasy, to make the subject matter clear.

Our friends from India Dr K L Kapoor (Hindu College, Delhi University) Prof S A Iqbal (Soifia P G College of Science and Education, Bhopal) and Dr Mohan Katyal St Stephen's College, Delhi University) have helped us a lot in compiling the matter and bringing it in this form. We are grateful to them.

We would most welcome the suggestions for its improvement.

September 1990

M Satake
T Nagahiro

CONTENTS

APPENDICES

A

Absorbance

The value of εlc in the expression

$$I = I_o \; 10^{-\varepsilon lc}$$

is known as absorbance (formely known as optical density or extinction). The symbol I_o and I stand for intensity of radiation before and after the absorption by a medium.

Absorption Coefficient

This coefficient was defined by Bunsen and is defined as follows.

The coefficient of absorption is the volume of gas reduced to STP (0 °C and 1 atm pressure) that has been dissolved by unit volume of solvent under a partial pressure of 1 atm of the gas.

Acidic Solution

If the concentration of H^+ in a solution is greater than its OH^- concentration, then the solution is said to be an acidic solution. In terms of K_w, the acid solution satisfies the expression

$$[H^+] > \sqrt{K_w} \text{ or } [OH^-] < \sqrt{K_w}$$

At 25 °C, the above expression becomes

$$[H^+] > 10^{-7} \text{ mol dm}^{-3} \quad \text{or} \quad [OH^-] < 10^{-7} \text{ mol dm}^{-3}$$

In terms of pH or pOH, the acidic solution satifies the expression

$$\text{pH} < 7 \quad \text{or} \quad \text{pOH} > 7 \qquad \text{at } 25\ ^\circ\text{C}$$

Acid

Various attempts have been made to define acids, starting from the phenomenological basis to the molecular composition.

The three definitions currently used are as follows.

The Arrhenius Concept: An acid is a substance which ionizes in water to produce $H^+(aq)$ or the hydronium ion.

The Bronsted—Lowry Concept: A substance is said to be an acid if it can donate a proton to another substance.

The Lewis Concept: An acid is a substance that can form a covalent bond by accepting an electron pair from some other substance.

Acid Base Indicators

Indicators, in general, are either organic weak acids or weak bases with a characteristic of having different colours in the ionized and unionized forms. For example, phenolphthalein is a weak acid (ionized form is pink and unionized form is colourless) and methyl orange is a weak base (ionized form is red and unionized form is yellow).

Activated Charcoal

The materials obtained by charring wood or coconut shells are able to take up large volumes of certain gases and he amounts can be increased by subjecting the charcoal to a process of activation; this generally involves heating in a vacuum or in inert gas or better, in steam, air, chlorine or carbon dioxide, at temperatures varying from 350 °C to 1000 °C.

Activated Complex Theory

The conversion of reactants to products (or vice versa) is through the formation of activated complex. This compound is not an intermediate molecule but is a molecule that is in the process of breaking or forming bonds. Eyring and others have developed the quantitative treatment of the activated complex theory (or transition state theory or absolute rate theory) based on the assumptions that there exists an equilibrium between the reactants and the activated complex and that the products are obtained by the decomposition of the activated complex, that is

$$A+B \longrightarrow {}^{\ddagger}X \longrightarrow \text{products}$$

where $X^{\ddagger}$ is the activated complex. Based on this theory, the expression of rate constant is

$$K=\frac{RT}{N_A h}\exp(-\triangle H^{o\ddagger}/RT)\exp(\triangle S^{o\ddagger}/R)$$

where $\triangle H^{o\ddagger}$ and $\triangle S^{o\ddagger}$ correspond to the equation

$$A+B\longrightarrow X^{\ddagger}$$

Activation Energy

The difference between the minimum energy required to bring about molecular rearrangement and the average energy of reactant molecules is known as activation energy.

Adiabatic Process

A process in which heat is allowed neither to enter nor to leave the system is known as adiabatic process.

Additivity Rules

The partial molar quantities are additive quantities. Thus

$$G=\sum_i n_i\mu_i$$

$$S=\sum_i n_i S_{i,pm}$$

$$V=\sum_i n_i V_{i,pm}$$

and so on.

Adiabatic Flame Temperature

Adiabatic flame temperature is a temperature which the system attains if the changes in system are carried out under adiabatic conditions. For example, if an exothermic reaction is carried out under adiabatic conditions, the heat evolved would raise the temperature of products. The final temperature is given by the expression

$$T_f=-\frac{\triangle H_{T_o}}{C_p(\text{products})}+T_o$$

where T_o is the temperature at which $\triangle H$ of the said change is stated and C_p is total heat capacity of the products.

If the reaction is carried out at constant volume, then the expression of T_f is

$$T_f = \frac{-\Delta U_{T_0}}{C_V(\text{products})} + T_0$$

Adsorbate

The substance on which adsorption takes place is known as adsorbate.

Adsorbent

The substance being adsorbed is known as adsorbent.

Adsorption

The adsorption implies the presence of excess concentration of any particular component at the surface of liquid or solid phase as compared to that present in the bulk of the material. This phenomenon is due to the presence of residual forces at the surface of the body.

The process of adsorption is an exothermic process.

Alkaline Solution

If the concentration of OH^- in a solution is greater than its concentration of H^+, the solution is said to be an alkaline solution. In terms of K_w, the alkaline solution satisfies the expression.

$$[OH^-] > \sqrt{K_w} \quad \text{or} \quad [H^+] < \sqrt{K_w}$$

at 25 °C, the above expression becomes

$$[OH^-] > 10^{-7} \text{ mol dm}^{-3} \quad \text{or} \quad [H^+] < 10^{-7} \text{ mol dm}^{-3}$$

In terms of pH or pOH, the alkaline solution satisfies the expression

$$\text{pOH} < 7 \quad \text{or} \quad \text{pH} > 7 \text{ at } 25\ °\text{C}$$

Alpha particles

The helium nucleus emitted in radioactive decays are known as alpha particles.

Amagat's Law of Partial Volumes

The total volume of a mixture of gases is equal to the sum of the partial volumes of the constituent gases. Mathematically, this law is written as

$$V_{total} = V_1 + V_2 + \ldots$$

where V_{total} is the total volume and V_1 and V_2 are the partial volumes. The lattter is defined as the volume which the individual gas would occupy if it were present alone in a container at temperature and pressure of the mixture.

Ampere

The ampere is that constant current which, if maintained in two straight parallel conductors of infinite length, of negligible cross section, and placed 1 metre apart in a vacuum, would produce between these conductors a force equal to 2×10^{-7} newton per metre of length.

The rate of current flow is expressed in amperes; 1 ampere is 1 coulomb of charge flowing in one second.

Amphiprotic Anion

An amphiprotic anion can either accept a proton from water or can donate a proton to water, i.e.

$$HA^- + H_2O \rightleftharpoons H_2A + OH^-$$

$$HA^- + H_2O \rightleftharpoons A^{2-} + H_3O^+$$

The pH of an aqueous solution of salt containing weak conjugate cation and an amphiprotic anion is given by the expression

$$pH = \tfrac{1}{2}(pK_{a1} + pK_{a2})$$

where K_{a1} and K_{a2} stand for

$$K_{a1} = \frac{[HA^-][H^+]}{[H_2A]}$$

and $$K_{a2} = \frac{[H^+][A^{2-}]}{[HA^-]}$$

Ampholyte

An electrolyte containing separate acidic and basic groups are known as ampholyte or amphoteric electrolyte. Example includes aminoacids (NH_2RCO_2H).

Amphoterism

The hydroxides of certain metals can function as both acids and bases. This phenomenon is known as amphoterism. The solubility of an amphoteric hydroxide is enhanced both in acidic as well as alkaline medium as compared to that in the neutral water.

Angle of Contact

The angle of contact of a liquid is defined as the angle which the side of a capillary tube makes with the tangent drawn at the meniscus touching the surface of the tube. This angle is measured within the liquid. For wetting liquids (such as water), the angle of contact is less than 90° and for non-wetting liquid (such as mercury), it is greater than 90°.

Angular Momentum and Magnetic Moment

The orbital motion and spinning of an electron produce angular moments which are given by the expressions

$$L=\sqrt{l(l+1)}\left(\frac{h}{2\pi}\right)$$

$$S=\sqrt{s(s+1)}\left(\frac{h}{2\pi}\right)$$

where l and s are the azimuthal and spin $(=\frac{1}{2})$ quantum numbers, respectively.

The negatively charged electron on orbiting around the nucleus and also spinning around its own axis produces magnetic dipole. The strength of magnetic dipole is determined from the magnetic moment value. The magnetic moments have quantized values and are given by the expressions

$$\textit{Orbital motion}\ \mu_m=-\mu_B\sqrt{l(l+1)}$$

$$\textit{Spin motion}\ \mu_m=-2\mu_B\sqrt{s(s+1)}$$

where μ_B, Bohr magneton, is given by the expression

$$\mu_B=\frac{eh}{4\pi m_e}$$

Its value is 9.274×10^{-24} J T^{-1}. The direction of magnetic moment vector is opposite to that of angular momentum.

Anharmonic Oscillator

For harmonic oscillation, potential energy goes on increasing with increase in the value of internuclear distance $\triangle x$ without showing any sign of breaking the molecule. In actual practise, the molecule breaks with increase in the value of $\triangle x$. The real molecule is said to execute anharmonic oscillatons as it breaks for the large value of $\triangle x$. The energy expression for the anharmonic oscillator is

$$E=(v+\tfrac{1}{2})\ h\nu\ [1-(v+\tfrac{1}{2})x_e]$$

where ν is the classical frequency of oscillation, x_e is anharmonicity constant and v is the vibrational quantum number: (0, 1, 2,...).

Anode

It is one of the terminals of a cell through which current leaves the cell. In the electrolytic cell, it is positively charged as it is attached to the positive end of the external battery. In the galvanic cell, it is negatively charged. The type of reaction taking place at this electrode is oxidation one.

Antibonding Molecular Orbital

If the energy of molecular orbital is larger than the energies of its constituent atomic orbitals, the molecular orbital is said to be an antibonding molecular orbital. The addition of electron in an antibonding molecular orbital destabilizes the molecule as it raises energy relative to the atoms taken separately.

Antifluorite Structure

In antifluorite structure, anions form cubical closest packing and cations occupy teterahedral holes in it.

Anti-Stokes Lines

Raman lines having higher energies than the energy of incident radiation are known as anti-Stokes lines. These lines correspond to the indirect demotion of the molecule from higher levels to low vibrational or/and rotational levels.

Antisymmetrical Stretching Vibration of Simple molecules

Stretching vibration involves the change in the internuclear distance. Antisymmetrical stretching vibration is observed when the outer atoms move in one direction and the central one in the opposite directions.

Antonoff's Rule

This rule states the value of interfacial tension (surface tension at the junction of two immiscible liquids). The rule is

$$\gamma_{AB}=(\gamma_{A}-\gamma_{B})$$

where γ_{AB} is the interfacial tension between two immiscible liquids A and B and γ_A and γ_B are the respective surface tensions.

Arrhenius Equation

The effect of temperature on rate constant, as given by Arrhenius, is

$$k=A\exp(-E_a/RT)$$

where A is a constant known as pre-exponential factor and E_a is known as activation energy. In the logarithm form, the expression is

$$\ln k=\ln A-\frac{E_a}{RT}$$

Thus, a plot of ln k versus $1/T$ is a straight line with slope equal to $-E_a/R$.

Arrhenius Equation to Calculate Degree of dissociation

The Arrhenius equation to calculate degree of dissociation of a weak electrolyte in a solution is

$$\alpha=\frac{\Lambda_e}{\Lambda\infty}$$

where Λ_e is the molecular conductivity at the given concentration and $\Lambda\infty$ is the corresponding value at infinite dilution where the weak electrolyte is present in the completely ionizied orm:

Arrhenius Theory of Dissociation

According to Arrhenius, a weak electrolyte in a solution exists in equilibrium with its characteristic ions. For example,

$$CH_3COOH \rightleftharpoons CH_3COO^- + H^+$$

$$NH_4OH \rightleftharpoons NH_4^+ + OH^-$$

Asymmetry Effect

Around each ion in a solution, there exists an oppositely charged ionic atmosphere. When the solution is subjected to potential difference across the two electrodes, centre ion moves in one direction whereas its ionic atmosphere in the opposite direction. This results in the distortion of ionic atmosphere. Consequently, the force exerted by the atmosphere on the central ion is no longer uniform and the ion experiences net force opposite to the direction of its motion resulting in the lowering of speed. This effect is known as asymmetry effect.

Atomic Mass

The average mass per atom of a specified isotopic composition of an element is known as atomic mass. It is given as

$$\text{mass of an atom}, A = A_r m_{au}$$

where A_r is the relative atomic mass and m_{au} is the atomic mass unit ($=1.6603\times10^{-27}$ kg)

The atomic mass is simply a mass and thus carries a unit of mass, i.e. (kg or g).

Atomic Mass Unit

The quantity (1/12)th of the mass of an atom ^{12}C is known as the atomic mass unit (symbol: m_{au}).

The value of atomic mass unit can be determined from the definition of the unit mole of the substance. One mole of the substance contains 6.023×10^{23} entities and for carbon=12, the mass of these number of atoms is 0.012 kg. Hence

$$m(^{12}C) = \frac{0.012 \text{ kg mol}^{-1}}{6.023\times10^{23} \text{ mol}^{-1}}$$

$$m_{au} = \frac{m(^{12}C)}{12} = \frac{1}{12}\left(\frac{0.012 \text{ kg mol}^{-1}}{6.023\times10^{23} \text{ mol}^{-1}}\right)$$

$$= 1.6603\times10^{-27} \text{ kg.}$$

Average Rate of Change of the Amount of a Reactant or a Product

By the term rate of change of amount of a reactant or a product, we mean the disappearance of the amount of a reactant or appearance of the amount of a product occurring in a unit interval of time. For a reaction

$$A \longrightarrow B$$

the average rates of change of the amounts over the time interval t_1 to t_2 are represented as

$$R_{av} = -\frac{\Delta n_A}{\Delta t} = -\frac{(n_A)_{t2} - (n_A)_{t1}}{t_2 - t_1}$$

$$R'_{av} = \frac{\Delta n_B}{\Delta t} = \frac{(n_B)_{t2} - (n_B)_{t1}}{t_2 - t_1}$$

For the given reaction, it is obvious that

$$R_{av} = R'_{av}$$

For the reaction

$$A \longrightarrow 2B$$

we will have

$$R_{av} = \tfrac{1}{2} R'_{av}$$

or $$2R_{av} = R'_{av}$$

where R and R' represent the rates of change of amount of of reactant and product, respectively.

The average rates of change over the same time interval decreases as the reaction proceeds.

Average Speed of Gaseous Molecules

The average of speeds possessed by molecules of a gaseous system is known an average speed. This is given by the expression

$$u_{av} = \sqrt{\frac{8RT}{\pi M}}$$

Aufbau Principle

Aufbau is a German word meaning building up. Thus, aufbau principle is the building up principle for writing the electronic configuration of various atoms. According to this principle, the electrons are fed into various orbitals in the increasing order of energy. This order is as follows.

$$1s < 2s < 2p < 3s < 3p < 4s < 3d < 4p < 5s < 4d < 5p < 6s < 4f \ldots$$

Avogadro's Law

Equal number of molecules of different gases under identical conditions of temperature and pressure occupy the same volume.

Avogadro Constant

This is equal to the number of particles in one mole of a substance. Its value is 6.023×10^{23} mol^{-1}.

Average Kinetic Energy

The average kinetic energy of gaseous molecules depends only on the temperature of the gas and is given by the expression

$$KE = (3/2)kT$$

where k is Boltzmann constant ($= 1.38 \times 10^{-23}$ J K^{-1}).

Azeotrope

Azeotrope is a Greek word meaning to boil unchanged. If the constitutents of a binary liquid solution exhibit very large deviations from Raoult's law, a maximum or minimum in the boiling point verses composition phase diagram is observed. At this point, both liquid and vapour curves meet each other, and thus both liquid and vapour have identical compositions. The system at this point boils at a constant temperature with the vapour phase having the same composition as that of liquid phase. The system at this point is known as azeotrope or azeotropic mixture.

Azimuthal Quantum Number

This number (symbol: l) describes the total angular momentum of the electron in an atom through the expression

$$L^2 = l(l+1)\left(\frac{h}{2\pi}\right)^2$$

The permitted values of l are 0, 1, 2,..., $(n-1)$ where n is the principal quantum number. It is customary to designate the values of l by letters as given below.

value of l	0	1	2	3	4	5
designation :	s	p	d	f	g	h

where the letters s, p, d and f are derived from the spectroscopy terms; sharp, principal, diffuse and fundamental, respectively.

The quantum number l dictates the shape of orbtial in an atom; s orbitals are spherical, p orbitals are dumb-bell shaped and other orbitals have more complicated shapes.

Autocatalyst

In some reactions, one of the products is able to catalyze the further decomposition of the reactants, such a substance is known as autocatalyst. One of the common examples is Mn^{2+} ions in the reaction between MnO_4^- and oxalate anion.

Balmes Series

See hydrogen spectrum.

Barometric Distribution Law

This law deals with the variation of pressure with height produced as a result of the influence of the gravitational field. The expression stating this law is

$$p = p_o \exp(-Mgh/RT)$$

where p is the pressure at height h, p_o is the pressure at zero height, g is acceleration due to gravity, M is the molar mass of the gas, T is its temperature and R is gas constant.

Bases

Various attempt have been made to define bases, starting from the phenomenological basis to the molecular composition. The three definitions currently used are as follows.

The Arrhenius Concept: A base is a substance which ionizes in water to produce $OH^-(aq)$.

The Bronsted—Lowry Concept: A substance is said to be a base if it can accept a proton from another substance.

The Lewis Concept: A base has an unshared electron pair which enables it to form a covalent bond with an atom, molecule or ion.

Bathochromic Shift

The general phenomenon of the shift of an absorption band to a region of longer wave length is often referred to as bathochromic shift.

Beattie Bridgeman Equation

This is the equation of state suggested for real gases. Equation is

$$p=(1-\gamma)\, RT\,(V_m+\beta)/V_m{}^2-\alpha V^2{}_m$$

where

$$\alpha=a_o\,(1+a/V_m)$$
$$\beta=b_o\,(1-b/V_m)$$
$$\gamma=c_o/V_mT^3$$

where a, b, a_o, b_o and c_o are constants.

Bending Mode of Simple Molecules

The bending mode involves the change in bond angle during the vibration of the molecule. In this mode, outer atoms move in opposite directions approximately at right angles to the molecular axis and the valence angle at the centre atom changes. This mode of vibration is normally doubly degenerate.

Berthelot Equation of State

This is an expression which is applicable to a real gas. Its form is

$$\left(p+\frac{n^2a}{TV^2}\right)(V-nb)=nRT$$

where p, V and T are the pressure, volume and temperature of the gas, n is the amount of gas and a and b are the constants characteristic of the gas.

BET Adsorption Equation

In order to account for the multilayer adsorption, Brunauer, Emmett and Teller derived the following expression for the adsorption.

$$V_{total}=\frac{V_{mono}\,C\,(p/p_\circ)}{(1-p/p_\circ)\,\{1+C\,(p/p_\circ)-(p/p^\circ)\}}$$

where

V_{total} is the total volume of gas adsorbed

V_{mono} corresponds to the monolayer adsorption

C is constant

p is the pressure of the gas

and $p_\circ$ is the vapour pressure of the liquefied adsorbent.

The above expression is generally written as

$$\frac{p}{V_{total}\,(p_\circ-p)}=\frac{1}{V_{mono}C}+\frac{C-1}{V_{mono}C}\;\frac{p}{p_\circ}$$

A plot of $p/\{V_{total}\,(p_\circ-p)\}$ against $p/p_\circ$ produces a straight line with intercept and slope as $1/(V_{mono}C)$, and $(C-1)/(V_{mono}C)$, respectivley.

Beta Particles

The electrons emitted in radioactive decays are known as beta particles.

Bohr Magneton

Bohr magneton is a basic unit of magnetic moment generated due to the electronic motions. It is given by the expression

$$\mu_B = \frac{eh}{4\pi m_e}$$

and has the value of 9.274×10^{-24} J T^{-1}.

Bohr Model of Atom

Bohr postulated the model of atom based on the hydrogen spectrum observed by Balmer. The basis postulates of his model are as follows.

1. The electron in an atom can revolve around the nucleus only in certain allowed circular orbits without losing any energy.
2. The electron can jump from one of the allowed orbits to another and can thereby gain or lose energy equivalent to the difference in energy of the two involved orbits.

The stationary orbits could be generated by imposing the quantum restriction on the angular momentum, i.e.

$$mvr = n\left(\frac{h}{2\pi}\right)$$

where n can have integral values, 1, 2, 3,..., and h is Planck's constant (6.626×10^{-34} J s)

The radius of allowed orbits is given by the expression

$$r = \frac{n^2}{Z} a_0$$

where $$a_0 = \frac{h^2}{4\pi^2 m\,(e/\sqrt{4\pi\varepsilon_0})^2}$$

where Z is the nuclear charge, m is the mass of electron, e is the electronic charge and ε_0 is the permittivity of the vacuum.

The energy of the electron in the allowed orbits is given by the expression.

$$E=-\frac{1}{n^2}\left(\frac{2\pi^2 mZ^2 (e/\sqrt{4\pi\varepsilon_0})^4}{h^4}\right)$$

Bohr Radius

The Bohr radius is defined as

$$a_0=\frac{h^2}{4\pi^2\, m(e/\sqrt{4\pi\varepsilon_0})^2}$$

and it is the distance between the first orbit in hydrogen atom from its nucleus. Its value is 52.9 pm.

Boyle's Law

At constant temperature, the volume of a definite mass of a gas is inversely proportional to its pressure, that is

$$V \propto 1/p$$

or $$V=K/p$$

where K is a constant of proportionality. According to this law, we can write

$$p_1V_1=p_2V_2 \qquad \text{(constant } T)$$

where V_1 and V_2 are volumes at pressures p_1 and p_2, respectively.

Boyle Temperature

At this temperature, a real gas behaves ideally over a wide range of pressure. This is due to the compensation of effects produced due to the size of molecules and intermolecular forces. The expression of Boyle temperature is

$$T_B=\frac{a}{Rb}$$

where R is Gas constant ($=8.314$ J K^{-1} mol^{-1}) and a and b are van der waals constants.

Binodal Curve

For a three-component system exhibiting one pair of partially miscible liquids, if a smooth curve passing through the points representing the compositions of conjugate solutions is drawn, a binodal curve is obtained.

Body-Centred Unit Cell

If in a unit cell, lattice point besides being at the corners is also present at the body centre of the unit cell, the resultant unit cell is known as body-centred unit cell. Only three body-centred unit cells, namely body-centred cubic, body-centred tetragonal and body-centred orthorhomic, are possible.

Boiling Point of a Liquid

The temperature at which vapour pressure of a liquid becomes equal the external pressure on the liquid is said to be the boiling point of the liquid. It the external pressure is 1 atm, the boiling point is known as normal boiling point.

Bond Dissociation Enthalpy

Bond dissociation enthalpy is defined as the enthalpy required to dissociate a given bond of some specific compound.

Bond Enthalpies

Bond enthalphy of a given bond is defined as the average of enthalpies required to dissociate the said bond present in different gaseous compounds into free gaseous atoms.

Bond Order

Bond order is defined as

$$\text{Bond order} = \frac{\text{Number of (bonding—antibonding) electrons}}{2}$$

This physical parameter gives the number of bonds between the two atoms of a molecule. Larger is the value of bond order, the more stable is the molecule and shorter is the internuclear distance between the two atoms.

Bonding Molecular Orbital

If the energy of molecular orbital is less than the energies of its constituent atomic orbitals, the molecular ortrital is said to be a bonding molecular orbtal. The addition of electron in a bonding molecular orbital stabilizes the molecule as it lowers energy and allows acculumation of electronic charge in the bonding regions.

Born Interpretation of the Wave function ψ

The square of the function when evaluated at a point is proportional to the propability of finding the electron at that point.

Bragg Equation

Bragg equation describes the essential condition for the constructive reflection of X-rays from the different sets of parallel planes of a crystal. Equation is

$$n\lambda = 2d \sin \theta$$

where n is the order of reflection and is equal to the number of wavelengths in the path difference between waves reflected by adjacent planes, d is the distance between the adjacent planes and θ is the angle of incidence of X-rays on the planes. Bragg equation is often written as

$$\lambda = 2\, d_{hkl} \sin \theta$$

where d is the perpendicular distance between adjacent planes having miller indices hkl.

Bravais Lattices

In seven crystal system, lattice points in unit cells are present at the corners. Such a unit cell is known as primitive unit cell. If the lattice points are also present at the centre of unit cell or at the centre of faces, the resultant unit cell is known as non-primitive unit cell. If all primitive and nonprimitive units cells are taken into account, a total of fourteen combinations are obtained. These are known as Bravais lattices. A brief account of these are as follows.

Triclinic	1	primitive
Monoclinic	2	;primitive, end-centred
Orthorhombic	4	;primitive, body-centred, end-centred, face-centred
Trigonal	1	primitive
Cubic	3	primitive, body-centred, face-centred
Tetragonal	2	primitive, body-centred
Hexagonal	1	primitive
Total	14	

Brownian Movement

The colloidal particles in water were found to be in constant motion with zig-zag path in all possible directions. This erratic motion is called the Brownian movement. It is due to the collisions of the colloidal particles with the molecules of the dispersion medium.

Buffer Capacity

Buffer capacity of a solution is defined as the change in the concentration of buffer acid (or base) required for changing its pH value by one, keeping $c_{acid}+c_{salt}$ or $c_{base}+c_{salt}$ constant. Mathematically, it is given as

$$\frac{\partial c(\text{HA})}{\partial \text{pH}} \text{ or } \frac{\partial c(\text{BOH})}{\partial \text{pH}}$$

The buffer capacity is maximum when the ratio of [salt]/[acid] (or [salt]/[base]) is one and it decreases as the ratio of [salt] to [acid] (or [salt] to [base]) changes in either direction.

Buffer Range

A buffer solution has useful buffer capacity provided the value of [salt]/[acid] (or [salt]/[base]) lies within the range of 10 to 0.1, i.e. ten parts of salt to one of acid (or base) at one extreme, to one part of salt to ten of acid (or base), at the other extreme. The pH of buffer range is pK_a-1 to pK_a+1 (or pK_b-1 to pK_b+1).

Buffer Solution

A solution which exhibits resistance to change its pH value with the addition of small quantity of acid or alkali is known as buffer solution. Such solutions usually consist of a mixture of a weak acid and salt of its conjugate base, e.g. acetic acid and sodium acetate (called acid buffers) or of a weak base and salt of its conjugate acid, e.g. ammonium hydroxide and ammonium chloride (called basic buffers). A salt of weak acid and a weak base, e.g. ammonium acetate also has a buffer action.

C

Cailletal and Mathias Law

This law is also known as law of the rectilinear diameter. The law states that the mean of the densities of any substance in a state of liquid and of saturated vapour at the same temperature, is a linear function of the temperature.

Calomel Half-Cell

This half-cell consists of mercury-mercurous chloride-chloride ion.

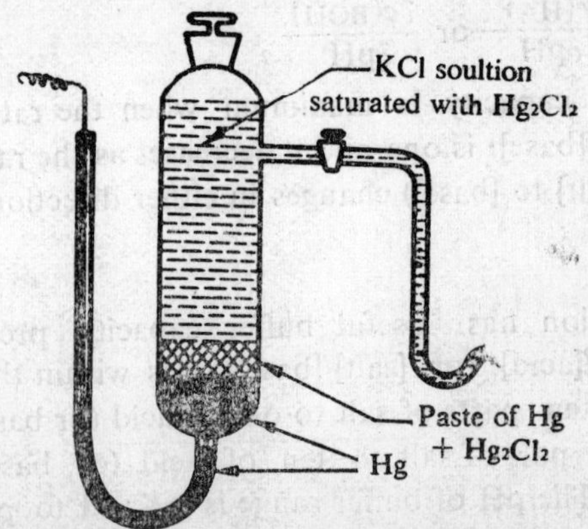

Capillary Action

The rise or fall of a liquid in a capillary tube is known as capillary action. This is due to the surface tension of the liquid.

Carnot Cycle

The Carnot cycle was named after Sodi Carnot, who was the first to describe this type of idealized reversible cycle. The system consisting of a single phase substance is subjected to the following reversible changes one after the other.

1. Isothermal reversible expansion from volume V_1 to V_2 at the higher temperature θ_2.
2. Adiabatic reversible expansion from volume V_2 to volume V_3 when temperature of the system is changed from θ_2 to θ_1.
3. Isothermal reversible compression from volume V_3 to volume V_4 at the temperature θ_1.
4. Adiabatic reversible compression from the volume V_4 to the original volume V_1, where temperature is also brought back to θ_2.

For a gaseous system, the above four steps leading to a cycle is shown in the following figure.

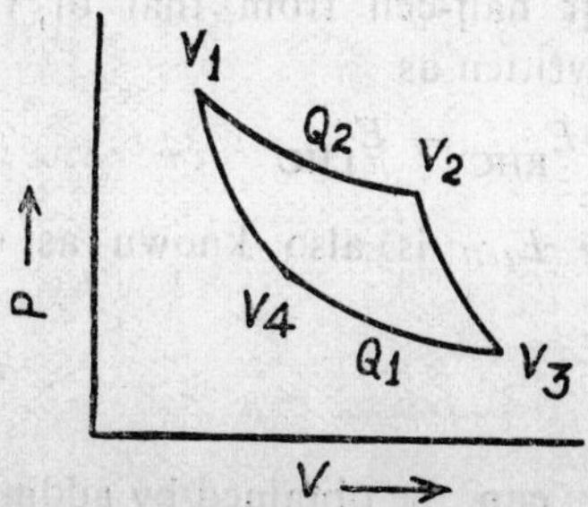

Catalyst

Sometimes, the rate of a chemical reaction is very much enhanced in the presence of a substance, called the catalyst. Though the catalyst is involved in the reaction, it does not appear in the overall reaction. In fact, a catalyst goes through a cycle in which it is used up and regenerated so that it is used over and over again.

Catalytic Poisons

In some cases, purely extraneous substances often present in very minute amounts are able to inhibit catalytic reactions; such substances are known as catalytic poisons.

Cathode

It is one of the terminals of a cell through which current enters into the cell. In the electrolytic cell, it is negatively charged as it is attached to the negative end of the external battery.

In the galvanic cell, it is positively charged. The type of reaction taking place at this electrode is the reduction one.

Cell Constant

The ratio l/A of the conductivity cell is known as cell constant, where l is the distance between the two electrodes and A is the area of cross section of the electrode.

Cell Potential

The cell potential can be obtained by subtracting half cell potential of left half-cell from that of right half-cell. It is conventionally written as

$$E_{cell} = E_{\text{RHC}} - E_{\text{LHC}}$$

The quantity E_{cell} is also known as electromotive force (emf) of a cell.

Cell Reaction

A cell reaction can be obtained by adding oxidation reaction of left half-cell (LHC) and reduction reaction of right half-cell

(RHC). Alternatively, it may be obtained by subtracting reduction reaction of LHC from that of RHC.

Chain Reactions

Chain reactions proceed through a complex sequence of elementary steps. The steps involved in chain reactions are

1. Chain initiation step
2. Chain propagation step
3. Chain inhibition step
4. Chain termination step.

An example of this type is provided by the reaction between H_2 and Br_2. Its steps are

Initiation : $Br_2 \xrightarrow{k_1} 2Br$

Propagation : $Br + H_2 \xrightarrow{k_2} HBr + H$

$H + Br_2 \xrightarrow{k_3} HBr + Br$

Inhibition : $H + HBr \xrightarrow{k_4} H_2 + Br$

Termination : $Br + Br \xrightarrow{k_5} Br_2$

Charles Law

The volume of a given mass of a gas at constant pressure is directly proportional to its absolute temperature, that is

$$V \propto T$$

or $$V = KT$$

where K is constant of proportionality. According to this law, we can write

$$\frac{V_1}{T_1} = \frac{V_2}{T_2} \qquad \text{(constant } p\text{)}$$

where V_1 and V_2 are the volumes at temperatures T_1 and T_2, respectively.

Chemical Actinometer

The number of photons absorbed by a photochemical system is determined through the use of chemical actinometer in which a photochemical reaction of known quantum yield is employed. The common actinometer is based on tne decomposition of oxalic acid, sensitized by uranyl ion in the wavelength region 250 to 440 nm. The reactions are

$$UO_2^{2+} + h\nu \longrightarrow (UO_2^{2+})^*$$

$$(UO_2^{2+})^* + (COOH)_2 \longrightarrow UO_2^{2+} + CO_2 + CO + H_2O$$

The amount of oxalic acid decomposed can be determined by titrating undecomposed oxalic acid against $KMnO_4$ solution. The quantum efficient of the above reaction is 0.57.

Chemical Potential

The chemical potential is the Gibbs function per mole of the substance, i.e.

$$\mu = \frac{G}{n}$$

For a multicomponent system with variable composition, the chemical potential is defined as

$$\mu_i = \left(\frac{\partial G}{\partial n_i} \right)_{T,\ p,\ n_j \atop j \neq i}$$

Chemical Shift

A particular type of magnetic active nuclei in a molecule may have different electron densities depending upon the atoms to which these are bonded. In the presence of a magnetic field, these electrons execute diamagnetic circulation around the nuclei. The diamagnetic effect produced depends upon the electron density. The actual magnetic field observed by a nucleus is given by the expression

$$B_{actual} = B_{applied} - B_{induced}$$

and consequently its potential energy in the magnetic field is given by the expression

$$V= -(\mu_N \; g \; m_J \;) B_{actual}$$

Thus a given type of magnetic active nuclei (say, proton) will have different potential energy depending upon the electron density around them. This facts leads to the different absorptions in the nuclear magnetic spectrum. The difference between the two absorption of the same type of magnetic active nuclei is known as chemical shift.

Chemisorption

If the forces of attraction between adsorbent and adsorbate are of chemical nature, the adsorption is said to be chemisorption. This adsorption has following characteristics:

1. usually occurs at high temperatures.
2. involves high heat of adsorption ($\simeq$80 to 420 kJ mol^{-1})
3. adsorption is slower
4. irreversible in nature
5. involves high activation energy
6. monolayer adsorption

Clausius Inequality

The statment of Clausius inequality is given below.

The energy of the universe is constant, the entropy of the universe always tends toward a maximum.

Clausius-Clapeyron Equation

This equation expresses the variation of vapour pressure of a condensed phase (solid or liquid) of a substance with change in temperature. The equation is

$$\log \frac{p}{1 \text{ atm}} = -\frac{\triangle H_m}{2.303\ RT} + I$$

where $\triangle H_m$ is the molar enthalpy ehange of phase transformation from condensed phase to the vapour phase and I is constant. According to this equation, a plot of log (p/1 atm) with $1/T$ is linear with scope equal to $-\triangle H_m/2.303R$. An alternate form of this equation is

$$\log \frac{p_2}{p_1} = -\frac{\triangle H_m}{2.303\,R}\left(\frac{1}{T_2} - \frac{1}{T_1}\right)$$

where p_2 and p_1 are the vapour pressures at temperature T_2 and T_1, respectively.

Clausius-Mosottii Equation

Clausius-mosottii equation is

$$\frac{(\varepsilon/\varepsilon_o)-1}{(\varepsilon/\varepsilon_o)+2}\,\frac{M}{\rho} = \frac{1}{3\varepsilon_o}\,N_A\,\alpha_d$$

where ε and ε_o are the permittivities of the dielectric and vaccum, respectively, M/ρ is the molar volume, N_A is Avogadro constant and α_d is the distortion polarizability. Clausius-Mosottii equation is applicable only for nonpolar molecules which do not possess permanent dipole moment.

Clausius Statement of Second Law of Thermodynamics

Clausius statement of second law of thermodynamics is as follows.

It is impossible for a cyclic process to covert heat into work without the simultaneous transfer of heat from a body at a higher temperature to one at a lower temperature or vice versa, i.e. it is impossible for a cyclic process to transfer heat from a body at a lower temperature to one at a higher temperature without the simultaneous conversion of work into heat.

Alternatively, it is impossible for an engine operating in a cycle to have as its only effect the transfer of a quantity of heat from a reservoir at a lower temperature to a reservoir at a higher temperature.

Classical Thermodynamics

The science which deals with the macroscopic properties of matter is known as classical thermodynamics. The entire formulation of thermodynamics can be developed without the knowledge that matter consists of atoms and molecules.

Coefficient of Compressibility

The coefficient of compressibility (symbol: κ) is defined as

$$\kappa=-\frac{1}{V}\left(\frac{\partial V}{\partial p}\right)_T$$

For an ideal gas

$$\kappa=\frac{1}{p}$$

Coefficient of Solubility

The coefficient of solubility was suggested by Ostwald and is defined as the volume of gas measured under given conditions of temperature and pressure that has been dissolved in a unit volume of the solvent.

Coefficient of Thermal Expansion

The coefficient of isothermal expansion (symbol: α) is defined as

$$\alpha=\frac{1}{V}\left(\frac{\partial V}{\partial T}\right)_p$$

For an ideal gas

$$\alpha=\frac{1}{T}$$

Coefficient of Viscosity

The coefficient of viscosity of a fluid is defined as the force per unit area required to maintain a velocity difference of unity between two parallel layers of fluid unit distance apart. The units of viscosity are

CGS units: dyn cm^{-2} s

SI units : N m^{-2} s

A unit of dyn cm^{-2} s is known as poise unit. Obviously,

1 poise$=10^{-1}$ N m^{-2} s

Colligative Properties

Colligative properties are those properties which depend upon the number of species present in the solution and not on their nature. These are:

1. relative lowering of vapour pressure,
2. depression in freezing point,
3. elevation of boiling point, and
4. osmotic pressure.

Collision Theory of Bimolecular Reactions

This theory is based on the following two postulates:

1. The products are formed only when the reactant molecules come close and collide with each other.
2. Only those collisions are effective in producing the products which satisfy the criterion of energy of activation and the specific orientation of molecules.

Based on these, it is possible to derive theoretical expression of rate constant. The expression is

$$k=pN_A \pi\sigma^2_{AB} \left(\frac{8kT}{\pi\mu}\right)^{\frac{1}{2}} \exp(-E_o/RT)$$

where p is steric factor which takes into account the specific orientation of molecules at the time of collision, σ_{AB} is the closeness of approach for the collisions and is equal to the sum of the radii of molecules A and B, i.e. $\sigma_{AB} = (\frac{1}{2} \ (\sigma_A + \sigma_B)$, μ is the reduced mass $(=m_A m_B \,/\, (m_A+m_B))$, N_A is Avogadro constant and E_o is the minimum energy to bring about the chemical reaction.

Colloids

Between the two types of solutions, namely, true solutions and suspensions, there are certain mixtures in which the dispersed particles are much bigger than the molecules or ions of a true solution but smaller than the particles in a suspension.

Mixture of this intermediate particle size are called colloids or sols.

The diameter of colloidal particles may range between 1 and 100 nm. The particles in the colloidal state do not settle down on standing, are not visible and can pass through a filter paper but not through parchment paper or animal membrane.

Common Ion Effect

The degree of dissociation of a weak electrolyte is suppressed in the presence of a common ion if present in the solution. For examples, the degree of dissociation of acetic acid is suppressed if the solution already contains H^+ or acetate anion, and the solubility of $BaSO_4$ in water is decreased if the solution already contains Ba^{2+} or SO_4^{2-} ions.

Compressibility Factor

It is the ratio of molar volume of a real gas and the molar volume ot an ideal gas, that is

$$Z = \frac{V_m(\text{real})}{V_m(\text{ideal})} = \frac{pV_m}{RT}$$

Conduction

The passage of electricity through a medium is known as conduction. The medium may be a metal, molten electrolyte and an aqueous solution of an electrolyte.

Conjugate Pair

An acid-base pair related through the loss or gain of a proton is called a conjugate pair. For example, NH_4^+ is the conjugate acid of the base NH_3 and NH_3 is the conjugate base of an acid NH_4^+. A strong acid produces weak conjugate base and vice verse.

Consolute Temperature

See critical solution temperature.

Conventional Value of Standard Molar Enthalpy

It is not possible to determine absolute value of enthalpy of a substance. However, a table of conventional values of standard molar enthalpies has been set up on the basis of the following reference.

The enthalpy of every element in its stable state of aggregation at 1.01325 bar (=1 atm) and 25 °C is assigned a zero value.

Coordinate-Covalent Bond

If the pair of electrons shared between two atoms comes exclusively from one of the atoms, the bond formed is said to be a coordinate covalent bond. One of the examples is $H_3N \rightarrow BF_3$.

Concurrent Elementary Reaction

If two or more products are obtained by the same reactant, such reactions are known as concurrent elementary reactions. Examples include the transformation of aldehydic form of D-glucose into α-D-glucose and β-D-glucose

Conductivity

Conductivity is the reciprocal of resistivity. This quantity may be considered as the conductance offered by a conductor of unit length and unit area of cross section. Its units are ohm^{-1} cm^{-1} or ohm^{-1} m^{-1}. The unit ohm^{-1} is also conventionally written as mho, which is ohm spelled backward. In SI units, ohm^{-1} is known as siemens (symbol: S).

Conductometric Titrations

The principle of conductometric titrations is based on the fact that during the titration, one of the ions is replaced by the other and invariably these two ions differ in the ionic conductivity with the result that the conductivity of the solution varies during the course of the titration. Figure illustrates the type of curves obtained for different types of conductometric titrations.

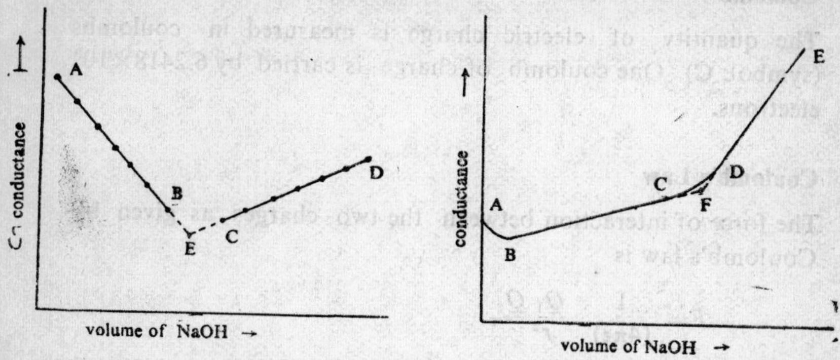

(a) Strong acid versus strong base. (b) Weak acid versus strog base.

Congruent Melting Point

A melting point where both solid and liquid of the same composition can coexist is known as congruent melting point.

Congruently Saturating Double Salt

If a double salt remains stable in the presence of water, it is said to be congruently saturating double salt.

Consecutive or Sequential Reactions

If the product formed in one of the elementary reactions acts as the reactant for same other elementary reaction, such reactions are known as consecutive or sequential reactions Examples include

$$CH_3OCH_3 \xrightarrow{k_1} CH_4 + HCHO$$

$$HCHO \xrightarrow{k_2} H_2 + CO$$

and $$(CH_2)_2O \xrightarrow{k_1} (CH_3CHO)^* \xrightarrow{k_2} CH_4 + CO$$

Corundum Structure

In corundum structure, anions form hexagonal closest packing and cations occupy only two-thirds of octahedral holes.

Coulomb

The quantity of electric charge is measured in coulombs (symbol: C). One coulomb of charge is carried by 6.2418×10^{18} electrons.

Coulomb's Law

The force of interaction between the two charges as given by Coulomb's law is

$$F = \frac{1}{(4\pi\varepsilon)} \frac{Q_1 Q_2}{r^2}$$

where r is ihe distance between the two charges Q_1 and Q_2 and ε is known as permittivity of the medium.

Coulometer

Coulometer is a device used in the measurement of quantity of electricity passed through a circuit. The silver coulometer is commonly employed for precise work. This coulometer consists of a platinum dish serving as both cathode and cell vessel and pure silver as anode. The electrolyte is an aqueous solution of purified silver nitrate. The coulometer is placed in the circuit in series. The mass of Ag deposited on the dish is determined when the experiment is over and from this mass, the quantity of electricity passed is determined through the expression

$$Q = \frac{m_{Ag} F}{(108 \text{ g mol}^{-1})}$$

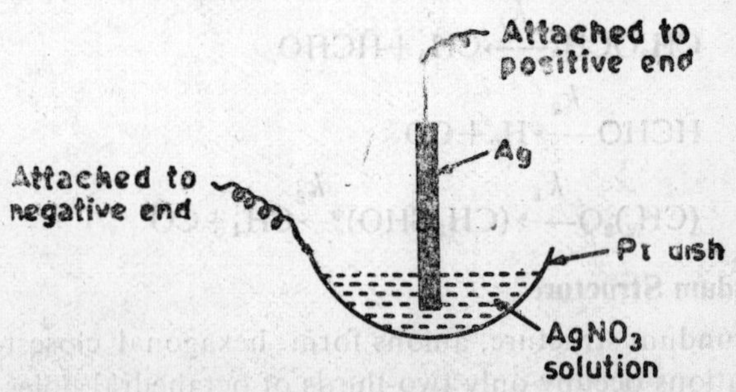

Covalent Bond

In the formation of a covalent bond, there occurs mutual sharing of electrons between the two involved atoms. This mechanism was suggested by GN Lewis.

In the formation of a covalent bond, each atom tries to acquire the stable configuration of eight electrons.

Craft's Rule

This rule is

$$\triangle T/°C = c\ (273.15 + t/°C)\ (760 - p/\text{mmHg})$$

where $\triangle T$ is the correction in boiling point t observed at pressure p and c is a constant having a value of 0.00012. For alcohols, water and carboxylic acids, c=0.00010 whereas for substances with very low boiling point (such as O_2, N_2, NH_3), c=0.00014.

Critical Pressure

It is the minimum pressure required to cause liquefaction of a gas at its critical temperature.

Critical Temperature

It is a maximum temperature at which a gas can be liquefied, i.e. the temperature above which a liquid cannot exist.

Critical Volume

It is the volume occupied by one mole of a gas at its critical temperature and critical pressure.

Critical Solution Temperature

For a partially miscible liquid pairs, a definite temperature at which or above which the two liquids are completely miscible in all proportion is known as critical solution (or consolute) temperature.

Critical Solution Temperature of a Three-Component System

At critical solution temperature or above this temperature, the three liquids if exhibit partial miscibility at lower temperature become completely miscible in all proportions.

Cryoscopic Constant

See depression in freezing point.

Crystal System

The morphological study of symmetry of unit cells has indicated that the crystals can be classified into seven crystal systems based on the presence of certain rotation axes. These are given below.

Crystal system	*Minimum symmetry of rotation*	*Unit cell dimensions**	*Examples*
Triclinic	1	$a \neq b \neq c$ $\alpha \neq \beta \neq \gamma \neq 90°$	$CuSO_4 . 5H_2O$ $K_2Cr_2O_7$
Monoclinic	2	$a \neq b \neq c$ $\alpha = \beta \neq 90° \neq \gamma$	S(monoclinic) $CaSO_4 . 2H_2O$ $Na_2SO_4 . 10H_2O$
Orthorhombic	222	$a \neq b \neq c$ $\alpha = \beta = \gamma = 90°$	S(rhombic) $BaSO_4$, KNO_3
Trigonal or Rhombohedral	3	$a = b = c$ $\alpha = \beta = \gamma \neq 90°$	$CaCO_3$, calcite
Cubic	Four 3	$a = b = c$ $\alpha = \beta = \gamma = 90°$	NaCl, diamond Alums, CaF_2
Tetragonal	4	$a = b \neq c$ $\alpha = \beta = \gamma = 90°$	TiO_2, Sn(white) $ZrSiO_4$
Hexagonal	6	$a = b \neq c$ $\alpha = \beta = 90°$, $\gamma = 120°$	SiO_2, graphite PbI_2, Mg, ZnO

*a unit cell is a parallelopiped having three sides (represented as *a*, *b*, and *c*) and three angles (represented as α, β and γ).

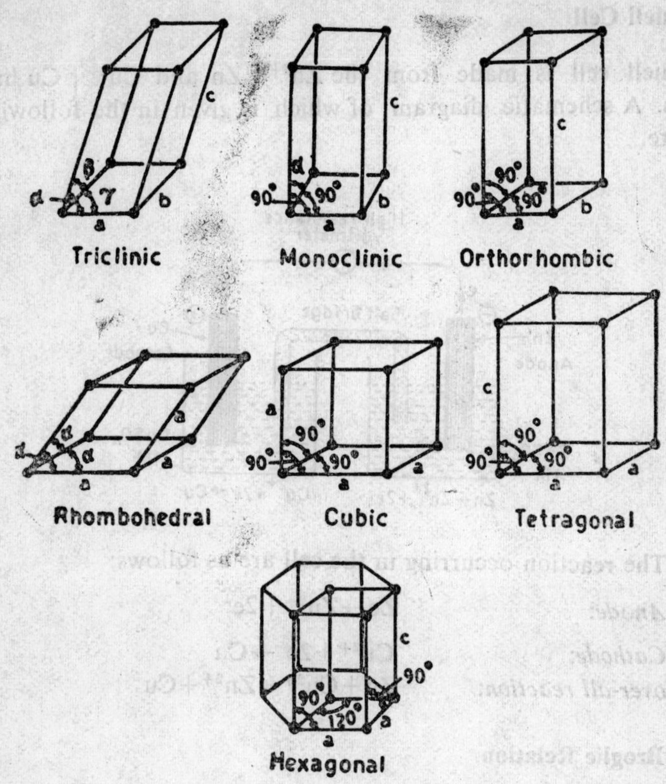

D

Dalton's Law of Partial Pressures

The total pressure of a mixture of noninteracting gases is equal to the sum of the partial pressures of the constituent gases, Mathematically, it is written as

$$p_{total} = p_1 + p_2 + \ldots$$

where p_{total} is total pressure and p_1 and p_2 are the partial pressures.

Daniell Cell

Daniell cell is made from the $Zn^{2+} | Zn$ and $Cu^{2+} | Cu$ half cells. A schematic diagram of which is given in the following figure.

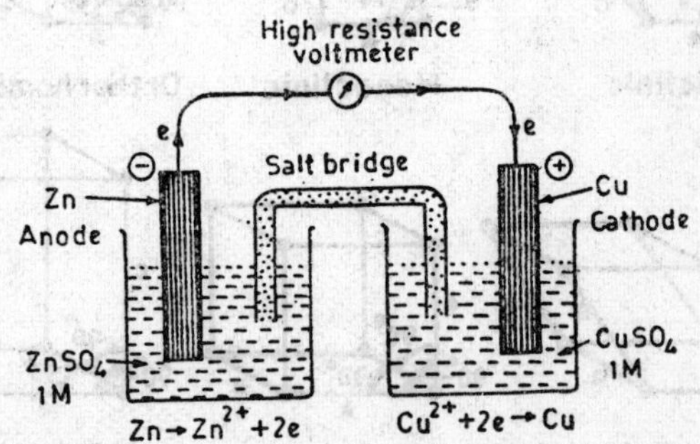

The reaction occurring in the cell are as follows:

Anode:	$Zn \rightarrow Zn^{2+} + 2e^-$
Cathode:	$Cu^{2+} + 2e^- \rightarrow Cu$
over-all reaction:	$Zn + Cu^{2+} \rightarrow Zn^{2+} + Cu$

De Broglie Relation

De Broglie relation is an expression which connects wave and particle nature of a substance. The relation is

$$\lambda = \frac{h}{p}$$

where λ is wavelength, h is Planck's constant and $p\ (=mv)$ is the linear momentum of the substance.

Debye Equation

Debye equation is

$$\frac{(\varepsilon/\varepsilon_0)-1}{(\varepsilon/\varepsilon_0)+2}\frac{M}{\rho} = \frac{1}{3\varepsilon_0} N_A \left(\alpha_d + \frac{p^2}{3kT} \right)$$

where

ε is the permittivity of the medium,
ε_0 is the permittivity of a vaccum,
M is the molar mass of dieletric in the medium,
ρ is the density of the medium,
N_A is Avogadro constant,
α_d is distortion polarizability,
p is dipole moment of the molecule,
k is Boltzmann constant,

and T is kelvin temperature.

The left side of the above expression is known as molar polarization (symbol: P_m).

Thus, a plot of P_m versus $1/T$ is a straight line with slope equal to $N_A\ p^2/9\varepsilon_0 k$. Knowing this value, the dipole moment of the molecule can be determined.

Debye-Falkenhagen Effect

When high frequency alternating current is used to measure conductivity of an electrolytic solution, its value is found to be higher than that determined at low frequency. This fact has been explained on the basis that at high frequency, the effect of ionic atmosphere on the conductivity disappears.

Debye Unit

See, dipole moment

Definition of pH

The hydrogen ion concentration of an aqueous solution can be conveniently expressed as pH$=$—log $\{[H^+]/\text{mol dm}^{-3}\}$.

Definition of pK_a

The equilibrium constant of a weak acid can be conveniently expressed as

$$pK_a = -\log K_a$$

Definition of pK_b

The equilibrium constant of a weak base can be conveniently expressed as

$$pK_b = -\log K_b$$

Definition of pK_w

The ionic product of water can be conveniently expressed as

$$pK_w = -\log (K_w/\text{mol}^2\ \text{dm}^{-6})$$

Since

$$K_w = [H^+]\ [OH^-]$$

we can write

$$-\log K_w = -\log [H^+] - \log [OH^-]$$

or $$pK_w = pH + pOH$$

i.e. the sum of pH and pOH of an aqueous solution is equal to the pK_w.

Definition of pOH

The hydroxyl ion concentration of an aqueous solution can be conveniently expressed as

$$pOH = -\log\{[OH^-]/\text{mol dm}^{-3}\}$$

Degenerate Orbitals

The orbitals having the same energy are known as degenerate orbitals. For examples, 2p orbitals in an atom are triply degenerate (corresponding to the three values of magnetic quantum number) and 3d orbitals are five-fold degenerate.

Degree of Dissociation

The degree of dissociation is equal to the fraction of the total substance present in the form of ions. This is usually represented by the symbol α. For example, for a weak electrolyte, if α is the degree of dissociation, we will have

$$AB \rightleftharpoons A^+ + B^-$$
$$c(1-\alpha) \quad c\alpha \quad c\alpha$$
$$A_2B \rightleftharpoons 2A^+ + B^{2-}$$
$$c(1-\alpha) \quad c(2\alpha) \quad c\alpha$$

where c is the total concentration of the weak electrolyte present in the solution

Deliquescence

If a salt is exposed to water vapour at a pressure greater than that of the saturated solution of its highest hydrate, at a given temperature, it will gradually take up water and finally form an unsaturated solution. This phenomenon is known deliquescence.

Depression in Freezing Point

When a nonvolatile solute is dessolved in a solvent, its freezing point is lowered. The relevant equation is

$$-\Delta T_f = K_f m$$

where K_f is the freezing point depression constant or the cryoscopic constant of the solvent and *m* is the molality of the solution. The units of K_f are K kg mol^{-1} and it is given by the expression

$$K_f = \frac{MRT_f^{*2}}{\Delta H_{fus}}$$

where M is molar mass solvent, $T_f{}^*$ is its freezing point and ΔH_{fus} is the molar enthalpy of fusion of the solvent.

Degree of Advancement of a Reaction

The progress of a reaction as dictated by its chemical equation can be conveniently expressed in terms of a physical quantity known as degree of advancement of reaction. This is represented by the symbol ξ. The degree of advancement when multiplied by the stoichiometric number of a substance appeared in the balanced chemical equation gives the change in the amount of that substance during the progress of the reaction. The unit of ξ is mol.

For example, for the decomposition of N_2O_5, we can write its chemical equation as

$$2N_2O_5 \longrightarrow 4NO_2 + O_2$$

If this reaction is studied with n_0 as the starting amount of N_2O_5, then we will have

$$2N_2O_5 \longrightarrow 4NO_2 + O_2$$

$t=0$ $\quad n_0 \quad 0 \quad 0$

$t=t$ $\quad n_0-2\xi \quad 4\xi \quad \xi$

where ξ is the degree of advancement of the reaction.

Degrees of Freedom

The degree of freedom or variance of the system is the minimum number of independent variables such as temperature, pressure and concentration, that must be ascertained so that a given system in equilibrium is completely defined.

Alternatively, the degree of freedom of the system may be defined as the number of factors, such as temperature, pressure and concentration, which can be varied independently without altering the number of phases.

Dialysis

Dialysis is a method in which colloidal particles are made free from excess of electrolytes to make them stable. A bag of parchment paper or cellulose acetate membrane is filled with the colloidal solution and is dipped into water. The ions diffuse through the membrane. The process can be speeded up by applying electric fied in the water outside the bag (electrodialysis).

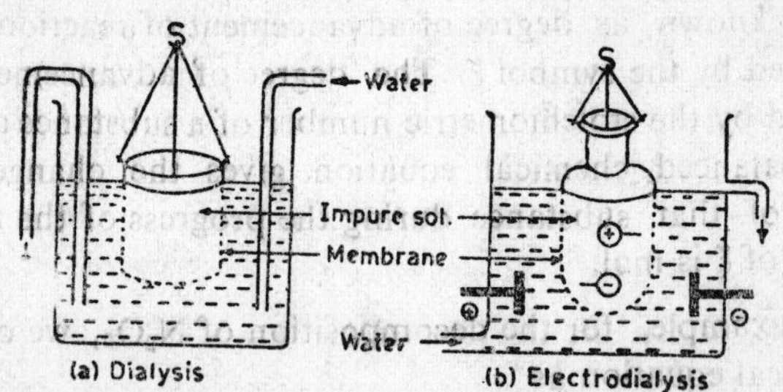

(a) Dialysis (b) Electrodialysis

Diamagnetism

The diamagnetism in atoms, ions or molecules is due to the orbital motion of electrons. A simple interpretation of diamagnetism is obtained if the orbital motion of electrons is considered to constitute a current flowing in a coil of wire. According to Lenz's law, when such a coil of wire is placed in a magnetic field, this induces the current so as to produce an induced magnetic field acting in the opposite direction of the applied magnetic field.

Diamagnetism is independent of temperature. It is exhibited by all types of substances whether the substance is diamagnetic or paramagnetic

Dielectric Constant

See, relative permittivity

Dieterici Equation of State

This is an expression which is applicable to real gases. Its form is

$$\{p \exp(na/VRT)\}\ (V—nb)=nRT$$

where p, V and T are the pressures, volume and temperature of a real gas ; n is the amount of the gas, and a and b are constants characteristic of the gas.

Differential enthalpy of Dilution

The differential enthalpy of dilution is defined as the enthalpy change when 1 mol of solvent is added to a large volume of the solution of known concentration so that there occurs no appreciable change in the concentration of the solution.

Differential Enthalpy of Solution

The differential enthalpy of solution may be defined as the change in $\triangle H$ per mol of solute when an infinitesimal amount of solute is added to a solution of definite concentration.

Another way of defining this quantity is as follows.

The differential enthalpy of solution is defined as the enthalpy change when 1 mol of solute is dissolved in a very large volume of a solution of known concentration so that there occurs no appreciable change in concentration of the solution.

Dipole Moment

The spatial distribution of bonding elcctrons between the two nuclei of a diatomic molecule depends on the nature of the two involved atoms. For homonuclear diatomic molecules such as H_2, N_2, etc. the distribution of electronic cloud is symmetrical around the two nuclei of the molecule. In heteronuclear diatomic molecules such as HCl, HBr, etc. the charge distribution is not symmetrical, the bonding electrons are more near to the more electronegative atom. Consequently, the molecule acquires separation of charges at the two ends, the more electronegative atom serves as the negative end and the positive end is the lesser electronegative atom. The distribution of charges in a molecule is described in terms of a physical parameter, known as the dipole moment. For diatomic molecules, it is defined as the product of charge at either end of the molecule and the distance between the two charges. It is a vector quantity having direction from negative to positivc cnd of the molecule. In a polyatomic molecule, each bond has a bond moment and the dipole moment of the molecule is the vector addition of these bond moments.

The unit of dipole mement is esu cm (CGS units) and C m (SI units). A unit of 10^{-18} esu cm is known as debye unit. Moreover $1D = 10^{-18}$ esu cm $= 3.3356 \times 10^{-30}$ C m.

Dissociation Constant of a Weak Acid

A weak acid in solution is present in equilibrium with its characteristic ions. For example, for acetic acid, we have

$$HA + H_2O \rightleftharpoons H_3O^+ + A^-$$

This equilibrium is characterized by its equilibrium constant, such that

$$K_{eq} = \frac{[H_3O^+]\,[A^-]}{[HA]\,[H_2O]}$$

Since water is present in excess, its concentration remains constant and can be merged with K_{eq} to give a new constant, known as dissociation or ionization constant of the acid. Thus, we have

$$K_a = \frac{[H_3O^+]\,[A^-]}{[HA]}$$

Dissociation Constant of a Weak Base

A weak base such as NH_4OH is present in equilibrium with its ions, such that

$$NH_4OH \rightleftharpoons NH_4^+ + OH^-$$

This equilibrium is characterized by its equilibrium constant, known as dissociation or ionization constant of the base, such that

$$K_b = \frac{[NH_4^+]\,[OH^-]}{[NH_4OH]}$$

Distribution Law or the Partition Law

If a solute is added in a system containing two immiscible liquids α and β (or slightly miscible liquids) and if the solute is soluble in in both the liquids, then it distributes itself between the two liquids in a definite manner, such that

$$\frac{c^{\alpha}}{c^{\beta}} = K_d$$

where c^{α} and c^{β} are the equilibrium molar concentrations of the solute in the two liquids and K_d is a constant known as the distribution coefficient or the partition coefficient. The value of the constant K_d depends only on the temperature of the system and is independent of the ralative amounts of the two layers and also that of the solute.

d-Orbital

The wave function having azimuthal quantum number equal to two is known as d-orbital.

Dot-Population Picture

In dot-population picture, the relative probability at a given location is shown by the density of dots near that location. For 1s and 2s orbitals, the dot-population pictures are given below.

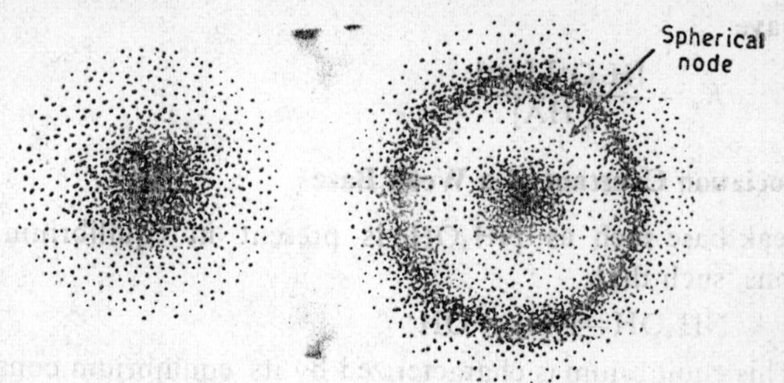

dot-population picture of 1s orbital

dot population picture of 2s orbital

Dry cell

A dry cell is represented as

$$Zn \mid NH_4Cl(20\%),\ ZnCl_2 \mid MnO_2 \mid C$$

The reactions involved are

Anode $Zn(s) \longrightarrow Zn^{2+} + 2e^-$

Cathode $2MnO_2(s) + H_2O(l) + 2e^- \longrightarrow Mn_2O_3(s) + 2OH^-$

Overall $Zn(s) + 2MnO_2(s) + H_2O(l) \longrightarrow Zn^{2+} + 2OH^- + Mn_2O_3(s)$

The OH^- generated causes the following secondary reactions.

$$NH_4Cl + OH^- \longrightarrow NH_3 + Cl^- + H_2O$$

$$Zn^{2+} + 2NH_3 + 2Cl^- \longrightarrow Zn(NH_3)_2Cl_2$$

The cell emf is 1.5 V.

Duhem-Margules Equation

A quantitative relation between the partial vapour pressures of the two constituents of a binary liquid system and their corresponding mole fraction is known as Duhem-Margules equation. The expression is

$$\frac{d \ln p_A}{d \ln x_A} = \frac{d \ln p_B}{d \ln x_B}$$

Dulong and Petit's Law

The product of the atomic mass and specific heat capacity of solid elements was found to be constant. This fact is known as Dulong and Petit's law

Dynamic Equilbrium

A dynamic equilibrium involves simultaneous conversions of reactants into products and products into reactants, the rate of conversion of these two processes are equal. All physical and chemical equilibrium involve dynamic equilibrium.

Ebullioscopic Constant

See elevation of boiling point.

Effect of Impurities on CST

If an impurity is soluble only in one of the two partially miscible liquids, its effect is to cause an increase in critical solution temperature (CST). If the impurity is soluble in both the liquids, it causes a decrease in CST. For example, the addition of NaCl increases the CST of phenol-water system and the addition of succinic acid decreases its CST.

Effect of Ionic strength on the Rate Constant of Ionic reactions

Based on the activated complex theory, it is possible to derive an expression

$$\log k = \log k_0 + z_A z_B \sqrt{\mu}$$

for the effect of ionic strength on the rate constant of ionic reaction in aqueous phase. Here

$$k_0 = K^{\ddagger}(RT/N_A h)$$

z_A and z_B are the charges on ions A and B, respectively, and μ is the ionic strength defined as

$$\mu=\tfrac{1}{2}\sum_i m_i z_i^2$$

where m_i is the molarity and z_i is the charge of the *i*th ionic species.

Effect of Pressure on Reaction Rate

Base on the activated complex theory, the expression of pressure dependence on reaction rate is

$$\ln k=-\frac{\triangle V^{\ddagger}}{RT}p+\text{constant}$$

wheee $\triangle V^{\ddagger}$ corresponds to the reaction

$$\text{Reactant}\longrightarrow X^{\ddagger}$$

where $X^{\ddagger}$ is the activated complex. The value of $\triangle V^{\ddagger}$ can be determined from the slope of a plot between ln k and p.

Effect of Pressure on vapour Pressure of a Liquid

The vapour pressure of a liquid (or solid) increases with the increase of external pressure. The expression is

$$RT\ln\frac{p_2}{p_1}=V_m(P_2-P_1)$$

where p_1 and p_2 are the vapour pressures at pressures P_1 and P_2, respectively.

Efficiency of Carnot Cyle

The efficiency of the cycle is defined as the ratio of the work done by the system to the amount of heat transferred to the system at the higher temperature. It can be shown that this efficiency is given as

$$\eta=1+\frac{q_1}{q_2}$$

where q_2 and q_1 are the heats involved in step 1 and 3, respectively. (*see*, Carnot cycle) The above expression is reducible to

$$\eta=1-\frac{\theta_1}{\theta_2}$$

where θ_2 and θ_1 are the temperatures in steps 1 and 2, respectively. Since $\theta_1<\theta_2$, $\eta<1$.

The following facts are observed for reversible Carnot Cycte.

1. Efficiency of Carnot Cycle is independent of the working substance.
2. Efficiency is independent of the size of the Cycle.
3. Efficiency of reversible cycle is greater than irreversible cyclic process.

Efforescence

The phenomenon of efforescence involves the loss of water by a hydrated salt. This is observed when the vapour pressure of a hydrate system is greater than that of water vapour in the air, and thus to attain equilibrium between the hydrate system and its surroundings, dehydration occurs.

Einstein of Energy

Energy carried by 1 mol of photons is conventionally known as one einstein of energy, i.e.

$$E=N_A\ h\nu$$

Electrochemical Cells

Electochemical cells can be broadly classified into two categories, viz. electrolytic cells and galvanic (or voltaic) cells. In the former, a chemical reaction (more precisely electrolysis) is carried out with the help of an electric current, whereas in the latter an electric current is produced as a result of some spontaneous chemical reaction.

Electrodialysis

See, dialysis.

Electrolytic Cell

An electrolytic cell is a device to convert electrical energy into chemical energy. With the help of electric current, electrolysis of salt is carried out. The characteristics of an electrolytic cell are as follows.

1. Cathode is negative electrode and reduction reaction takes place at this electrode.
2. Anode is positive electrode and oxidation reaction takes place at this electrode.
3. Electrons from the negative end of the external battery are forced into the electrolytic cell through the cathode.
4. Electrons are taken out of the electrolytic cell from the anode and are given back to the positive end of the external battery.

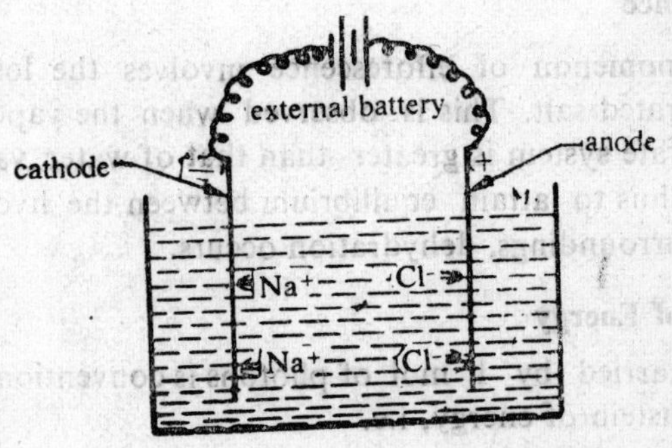

Electromotive Eorce

The potential difference between the two electrodes of a galvanic cell is known as electromotive force. It is given by the expression

$$E_{cell} = E_{RHC} - E_{LHC}$$

where E_{RHC} and E_{LHC} are the reduction potentials of right hand cell and left hand cell, respectively.

Electrophoresis

The existence of electric charge on the colloidal particles can be exhibited by the phenomenon of electrophoresis. It involves the motion of colloidal particles either towards the cathode or anode under the influence of an electric field.

Electron Spin Resonance Spectroscopy

This branch of spectroscopy deals with the change of spin of an electron in the presence of a magnetic field. The spinning of electron produces a tiny dipole whose magnetic moment is given by the expression

$$\mu_m = -2\mu_B \sqrt{s(s+1)}$$

where μ_B is Bohr magneton ($=9.2741 \times 10^{-24}$ J T^{-1}) and $s=\frac{1}{2}$. Its potential energy in the presence of a magnetic field is

$$V = 2\mu_B \, m_s B$$

where m_s is $+\frac{1}{2}$ or $-\frac{1}{2}$. The difference of potential energy between α and β levels is

$$\Delta V = 2\mu_B \, B$$

For $B \approx 0.3$ T, ΔV lies in the radiofrequency region. The absorption of appropriate frequency in this region produces electron spin resonance spectrum.

Electrophoretic Effect

Ions in solutions are generally solvated and when these move, they carry with them the associated solvent molecules. Thus, the central ion moves in a medium in which solvent molecules carried by the ionic atmosphere move in the opposite direction. Thus, the movement of ions is not in a stationary medium but through a medium which moves in opposite direction. This counter currents slow down the speeds of ions. This effect is known as electrophoretic effect.

Elementary Step

An elementary step is the one in which reactants are converted into products in a single step. The elementary reactions are classified according to the number of molecules which they involve. A process in which only one molecule is involved is known as unimolecular process. One involving two molecules is called bimolecular and so on. An elementary step with molecularity greater than three are not known. This follows from the fact that reactants are converted into products when the reactant molecule colloide together one at the same place, and molecular collisions involving more than three molecules are not known. The order of elementary step is equal to its molecularity.

Elevation of Boiling Point

When a nonvolatile solute is dissolved in a solvent, its boiling point is raised. The relevant equation is

$$\triangle T_b = K_b m$$

where K_b is the boiling point elevation constant (or ebullioscopic constant) and m is molality of the solution. The units of K_b are K kg mol^{-1} and it is given by the expression

$$K_b = \frac{MRT_b^{*2}}{\triangle H_{m,vap}}$$

where M is the molar mass solvent, T_b^* is its boiling point and $\triangle H_{m,vap}$ is the molar entalpy of vapourization of the solvent.

Eigenvalue Problem

A differential cquation of the type

$$H_{op}\psi = E\psi$$

is said to be an eigenvalue problem, where the operator H_{op} when acts on the function ψ gives back the function ψ multiplied by a constant E. The function ψ is known as eigen function and the constant E is known as eigenvalue.

Enantiotropy

If the change of a substance from one polymorphic form to another can be reversed, the substance is said to exhibit enantiotropy. The common example is rhombic sulphur to monoclinic sulphur and vice versa.

End-Centred Unit Cell

If in a unit cell, lattce points besides being at the corners are also present in the centre of two equivalent faces, the resultant unit cell is known as end-centred unit cell.

Only two end-centred unit cells, namely, end-centred orthorhombic and end-centred monoclinic, are possible.

Endergonic Reactions

Reactions for which $\triangle G > 0$ are called endergonic reactions.

Endothermic Reactions

The reactions which are brought about by the absorption of heat are known as endothermic reactions.

Enthalpy Function

Like internal energy, the enthalpy of a system is defined as

$$dH = dq_p$$

or

$$\triangle H = q_p$$

that is, heat transferred at constant pressure changes the enthalpy of the system.

Enthalpy of Combustion

Enthalpy of combustion of a given compound is defined as the enthalpy change when one mole of this compound combines with the requisite amount of oxygen to give products in their stable forms. For example, the enthalpy of combustion of methane at 25 °C is −890.36 kJ mol^{-1}. This implies the following equation.

$$CH_4(g) + 2O_2(g) \longrightarrow CO_2(g) + 2H_2O(l)$$

$$\triangle H^\circ_{298K} = -820.36 \text{ kJ mol}^{-1}$$

Enthalpy of Formation

The enthalpy of formation of a given compound is defined as the enthalpy change when one mole of a given compound is formed starting from the elements in their stable states of aggregation. For example, enthalpy of formation of cane sugar

($C_{12}H_{22}O_{11}$) is −2218 kJ mol^{-1}. This implies the following chemical equation.

$$12C(s)+11H_2(g)+\tfrac{11}{2}O_2(g)\longrightarrow C_{12}H_{22}O_{11}(s)$$
$$\triangle H^\circ=-2218 \text{ kJ mol}^{-1}$$

Enthalpy of Hydration

Enthalpy of hydration of a given anhydrous or partially hydrated salt is the enthalpy change when it combines with the requisite amount of water to form a new hydrated stable salt.

Enthalpy of Ionization

The enthalpy of ionization is defined as the enthalpy change when one mole of a weak acid (or base) in dilute solution is made to ionize to produce the ions). For example, the enthalpy of ionization of HCN is −12.13 kJ mol^{-1}. This implies the following cquation.

$$HCN \longrightarrow H^+ + CN^-$$
$$\triangle H=-12.13 \text{ kJ mol}^{-1}$$

Enthalpy of Neutralization

Enthalpy of neutralization is defined as the enthalpy change when one mole of H^+ in dilute solution combines with one mole of OH^- to give rise to undissociated water, i.e.

$$H^+(aq)+OH^-(aq)\longrightarrow H_2O(l)$$
$$\triangle H_{neut} \simeq -57.3 \text{ kJ mol}^{-1}$$

Enthalpy Precipitation

Enthalpy of precipitation is the enthalpy change when one mole of a precipitate is formed.

Enthalpy of Transition

Enthalpy of transition is the enthalpy change when one mole of one allotropic form changes to another.

Entropy and Probability

By using statistical mechanics, it can be shown that

$$S=k \ln W$$

where k is Boltzmann constant and W is the number of mioro-states associated with the predominant configuration of the system.

Entropy Change in a Chemical Reaction

The entropy change for a chemical equation can be determined using third-law entropy by using the expression

$$\Delta S^\circ = \sum_i \nu_i S_i^\circ(\text{product}) - \sum_i \nu_i S_i^\circ(\text{reactant})$$

where ν_i is the corresponding stoichiometric number appeared in the balanced chemical equation.

Entropy Change in a reversible Phase Transformation

Since the reversible phase transformation takes place a constant equilibirum temperature, it is obvious that

$$\Delta S = \frac{q_{rev}}{T}$$

where q_{rev} is the heat involved in the plase transformation. Since

$$q_{rev} = \Delta H_m$$

we have

$$\Delta S_m = \frac{\Delta H_m}{T}$$

Entropy Function

The entropy function, defined in terms of its defferential, is

$$dS = \frac{dq_{rev}}{T}$$

The unit of entropy function is J K^{-1}

The entropy function is a function of the independent variables which are used to define the state of a system. It is an extensive function. The change in the value of the entropy function in going from one state to another is independent of the path and the cyclic intergral of dS for a cyclic change of state is always zero.

The entropy function is a measure of disorderliness of the system—larger the disorderliness, larger is the entropy of the system. Take, for example, the three states of a substance, namely, the solid, liquid and gaseous states. In general, the molecules in the gaseous states are more disordered than those in the liquid state, while the molecules in the latter are more disordered than those in the solid state. Thus, the entropy of the substance in these three states of matter follows the order.

$$S(\text{gaseous state}) >> S(\text{liquid state}) > S(\text{solid state})$$

Entropy of mixing

If two or more gases at the same T and p are mixed together, the entropy change in the mixing process is given by the expression

$$\triangle S_{mix} = -nR \sum_i x_i \ln x_i$$

where x_i is the mole fraction of the constituent i in the mixture and n is the total amount of gases in the mixture. Since $x_i < 1$, $\triangle S_{mix}$ has a positive value indicating that mixing of gases is a spontaneous process.

Enzyme Catalysis

One of the most important examples of homogeneous catalysis is the catalysis of reactions in biological systems by enzymes. Enzymes are complex protein molecules and are usually very specific, catalyzing only one type of reaction.

Eotvos Equation

Eotvos equation is

$$-\frac{\mathrm{d}[\gamma (M_\mathrm{v})^{2/3}]}{\mathrm{d}T} = k$$

that, the rate of decrease of molar surface energy of a liquid with temperature has a constant value. The integrated form of the above expression is

$$-\frac{\gamma_1 (M_{\mathrm{v}_2})^{2/3} - \gamma_2 (M_{\mathrm{v}_2})^{2/3}}{t_1 - t_2} = k$$

At critical temperature, where molar surface energy has zero value, the above expression becomes.

$$\gamma\ (M\nu)^{2/3}=k(t_c-t)$$

Equal Probability Contours

In equal probability contours, the contours by joining the points of identical probability are drawn. If we are contended with a total of 90 per cent probability (a fairly large probability) of finding the electron, we can can draw a contour within which there exists a total of 90 per cent probability of finding the electron. This gives definite shape in three dimensions and is known as the shape of the orbital.

Equation of State

An expression which represents the relationship between pressure, volume and temperature of a given mass of a gas is known as equation of state. A few examples are

Ideal Gas: $pV=nRT$

Van der Waals Gas: $\left(p+n^2\frac{a}{V^2}\right)\left(V-nb\right)=nRT$

(where a and b are van der Waals constants. These are characteristic of the gas.)

Equilibrium Constant, K_c

The dynamic equilibrium present between unionized electrolye and its ion can be characterized by a constant, known as equilibrium constant. This is defined as

$$K_{eq}=\frac{\text{Multiplication of concentration of products, each raised to a power equal to the corresponding stoichiometric number}}{\text{Multiplication of concentration of reactants, each raised to a power equal to the corresponding stoichiometric number}}$$

According to IUPAC (International Union of Pure and Applied chemistry), a chemical equation can be written as

$$0=\sum_i \nu_i B_i$$

where ν_i are positive for products and are negative for reactants, and B_i are different species (reactants and products) of the reaction.

The expression of equilibrium constant is written as

$$K_{eq}=\prod_i (B_i)^{\nu_i}$$

where the sign Π stands for multiplication.

Equilibrium Constant, K_p

If the reaction is at equilibrium, the reaction quotient of its chemical equation is known as equilibrium constant (symbol: K_p). Its definition is

$$K_p=\prod_i \left(p_i \right)_{eq}^{\nu_i}$$

where ν_i are the stoichiometric numbers which have positive values for the products and negative values for the reactants. The subscript *eq* stands for equilibrium pressure. The sympol Π stands for multiplication sign.

If each pressure is divided by the standard unit pressure, one gets the standard equilibrium constant (symbol: $K_p°$). It is defined as

$$K_p°=\prod_i \left(\frac{p_i}{p°} \right)_{eq}^{\nu_i}$$

where $p°$ is 1 atm. Taking an example of

$$2N_2O_5 \rightleftharpoons 4NO_2+O_2$$

we have

$$K°_p=\frac{\left\{p(NO_2)/p°\right\}_{eq}^{4}\left\{p(O_2)/p°\right\}_{eq}}{\left(p(N_2O_5)/p°\right)^2}$$

Equivalent Conductivity

The equivalent conductivity (symbol: Λ_{eq}) of an electrolyte may be defined as the conductance of a volume of solution contanining one equivalent mass of a dissolved substance when placed between two parallel electrodes which are at a unit distance apart, and large enough to contain between them the whole solution. Mathematically, it is defined as

$$\Lambda_{eq}=\frac{\kappa}{c}$$

where κ is the conductivity and c is the equivalent concentration. The units of Λ_{eq} are Ω^{-1} cm^2 eq^{-1}.

Escaping Tendency

GN Lewis proposed the term escaping tendency for the chemical potential (μ).

A given substance can be transferred spontaneously from region α to the region β if $\mu^{\alpha} > \mu^{\beta}$. The flow will continue till $\mu^{\alpha} = \mu^{\beta}$.

Euler's Resiprocity Relation

This relation expresses the necessary condition for a differential $d\phi$ given as

$$d\phi=\left(\frac{\partial\phi}{\partial x}\right)_y dx+\left(\frac{\partial\phi}{\partial y}\right)_x dy$$

to behave as an exect differential. The relation is

$$\left[\frac{\partial}{\partial y}\left(\frac{\partial\phi}{\partial x}\right)_y\right]_x=\left[\frac{\partial}{\partial x}\left(\frac{\partial\phi}{\partial y}\right)_x\right]_y$$

Euler's Theorem

If a function is homogeneous of degree n, according to Euler's theorem, it should satisfy the following relation

$$\left(\frac{\partial f}{\partial x}\right)_{y,z\cdots}+y\left(\frac{\partial f}{\partial y}\right)_{x,z,\cdots}+\ldots=nf(x,y,\ldots)$$

Eutectic Point

If the composition of crystallizing solid mixture at the eutectic temperature has the same value as that of liquid phase from which crystallization takes place, the point representing this situation at the eutectic temperature is known as eutectic point. The composition of solid or liquid phase at the eutectic point is known as eutectic composition.

Eutectic Temperature

If in a binary liquid system, only pure components crystallize on cooling, then it exhibits a lowest temperature at which both components crystallize together from a liquid solution of any composition. This temperature is known as eutectic temperature. Alternatively, if may be defined as the temperature at which a solid mixture of two components of any composition starts melting.

Evaluation of $\triangle H$ due to the Temperature Change

The expression of dH is

$$dH = dq_p$$
$$= C_p \, dT$$

where C_p is the heat capacity of the system at constant pressure. For a finite change, assuming C_p constant, we have

$$\triangle H = C_p \triangle T$$

Heat capacity at constant pressure is temperature dependent and the expression relating these two is

$$C_p = a + bT + cT^2 + \ldots$$

where a, b and c are are constant. If this equation is employed, them $\triangle H$ is given by the expression

$$\triangle H = a\,(T_2 - T_1) + \frac{b}{2}\,(T_2^2 - T_1^2) + \frac{c}{3}\,(T_2^3 - T_1^3) + \ldots$$

Evaluation of $\triangle U$ due to the Temperature Change

The expression of dU is

$$dU = dq_V$$
$$= C_V dT$$

where C_V is the heat capacity of the system at constant volume. For a finite change, assuming C_V constant. we have

$$\triangle U = C_V \triangle T$$

Exact differential

In the differential expression

$$d\phi = P_{x,y}\, dx + Q_{x,y}\, dy$$

if

$$\left(\frac{\partial Q}{\partial x}\right)_y = \left(\frac{\partial P}{\partial y}\right)_x$$

then the differential $d\phi$ is known as exact differential.

Exergonic Reaction

Reactions for which $\triangle G < 0$ are called exergonic reactions.

Exothermic Reaction

The reactions which are accompanied by the release of heat are known as exothermic reactions.

Expression Relating C_p and C_V

First law of thermodynamics helps in establishing the relation connecting C_p and C_V. The expression is

$$C_p - C_V = \left[p + \left(\frac{\partial U}{\partial V}\right)_T \right] \left(\frac{\partial V}{\partial T}\right)_p$$

or $$C_p - C_V = \left[V - \left(\frac{\partial H}{\partial p}\right)_T \right] \left(\frac{\partial p}{\partial T}\right)_V$$

or $$C_p - C_V = \left[V - \left(\frac{\partial H}{\partial T}\right)_p \left(\frac{\partial T}{\partial p}\right)_H \right] \left(\frac{\partial p}{\partial T}\right)_V$$

where the various symbols have their usual meanings.

For an ideal gas, the right hand side of the above expressions is equal to nR, where n is the amount of gas and R is gas constant. Hence,

$$C_p - C_V = nR$$

Extensive Variables

The properties which are dependent on the mass of the system are known as extensive variables. Examples are: volume, energy, heat capacity, enthalpy, entropy, free energy, length and mass.

Extent of Reaction

The progress of a reaction as dictated by its chemical equation can be conveniently expressed in terms of a physical quantity known as extent of reaction. This is represented by the symbol *x*. The extent of reaction when multiplied by the stoichiometric number of a substance appeared in the balanced chemical equation gives the change in concentration of that substance during the progress of the reaction. The units of extent of reaction is mol per unit volume, e.g. mol L^{-1}.

For example, for the decomposition of N_2O_5, we can write its chemical equation as

$$2N_2O_5 \longrightarrow 4NO_2 + O_2$$

If the reaction is studied with $[N_2O_5]_0$ as the starting concentratoin of N_2O_5, then we will have

	$2N_2O_5 \longrightarrow$	$4NO_2+$	O_2
$t=0$	$[N_2O_5]_0$	0	0
$t=t$	$[N_2O_5]_0-2x$	$4x$	x

where *x* is the extent of reaction.

Face-Centred Unit Cell

If in a unit cell, lattice points besides being at the corners are also present in the centre of all the six faces, then the resultant unit cell is known as face-centred unit cell. Only two face-centred unit cells, namely face-centred cubic and face-centred orthorhombic, are possible.

Faraday Constant

Faraday constant is the total charge carried by 1 mol of electrons. Its value is 96487 C mol^{-1}.

Faraday's Law of Electrolysis

The quantitative relationships between electricity and chemical change were described by Michael Faraday. These are:

1. The mass of a chemical substance involved at an electrode is directly proportional to the amount of current passed through the cell.
2. The masses of different substances produced by a given amount of current are proportional to the equivalent masses of the substances.

Mathematically, the above two laws are expressed as

$$m = \frac{Q}{F} \frac{M}{z}$$

where Q is the quantity of electricity passed, F is Faraday constant (=96487 C mol^{-1}), M is the molar mass and z is the charge number of the cation or anion involved.

Fick's Law of Diffusion

Fick's law of diffusion is

$$dw = -D \frac{dc}{dx} dt$$

where dw is the quantity of the substance diffusing across 1 cm^2 in time dt when the concentration gradient is $-dc/dx$, and D is the diffusion coefficient. The latter is equal to mass of material diffusing across a plane of 1 cm^2 in area in unit time under a concentration gradient of unity.

First Law of Crystallography

The angles between the corresponding faces of various crystals of the same substance are constant. This statement is known as first law of crystallography.

First Law of Thermodynamic

This law is essentially the law of conservation of energy, according to which, the internal energy of an isolated system remains constant irrespective of what changes of state may occur in an isolated system.

Mathematically, this law is written as

$$dU = dq + dw$$

or $$\triangle U = q + w$$

where dU, dq and dw stand for the infinitesimal changes in internal energy, heat and work, respectively. Symbols $\triangle U$, q and w are the corresponding values for a finite process.

First Order Phase Transition

In the first order phase transition, changes are observed in the values of enthalpy and volume during the transition, i.e. $\triangle H$ and $\triangle V$ are nonzero.

The other characteristics of this transition are as follows:

1. The chemical potential is a continuous function of temperature.
2. The derivatives of chemical potential are discontinuous.
3. System has infinite heat capacity at the transition point.

The ordinary phase transitions are the examples of this type of transitions.

f Orbital

The wave function having azimuthal quantum number equal to three is known as f orbital.

Fluidity

Fluidity is an inverse of viscosity, that is

$$\phi = \frac{1}{\eta}$$

Fluorescent Spectrum

If the emission electronic spectrum is slightly shifted to the longer wavelength (red shift) as compared to the absorption spectrum the spectrum is said to be fluorescent spectrum. What happens that in the excited state, the excess vibrational energy is transferred to the neighbouring molecules during the internuclear collisions. This transference appears as the kinetic energy and hence heating effects are produced. The transfer of energy in molecular collisions is termed as radiationless decay.

Fluorite Structure

In fluorite structure, cations form cubical-closest packing anions occupy tetrahedral holes in it.

Franck-Condon Principle

The electronic transition occurs so quickly that the internuclear separation during the transition remains unchanged.

Thus, the instantaneous internuclear separation in the excited electronic state is the same as it was in the initial ground state at the time of electronic transition. As a consequence of this, the electronic transitions are shown by vertical lines in the energy levels depicting ground and excited electronic states.

Free Energy Change of a Cell Reaction

Free energy of a cell reaction is related to the cell potential by an expression

$$\triangle G = -nFE_{cell}$$

Free Energy of Mixing

The free energy of mixing during the mixing of different ideal gases at the same T and p is given by the expression

$$\triangle G_{mix} = nRT \sum_i x_i \ln x_i$$

where x_i is the mole fraction of the constituent i and n is the total amount of gases in the mixture. Since $x_i < 1$, $\triangle G_{mix}$ has a negative value indicating that mixing is a spontaneous process.

Freezing Point of a Liquid

The temperature at which vapour pressure of a liquid becomes equal to that of its solid phase when the external pressure is 1 atm is known as freezing point of a liquid.

Freundlich Adsorption Equation

An empirical expression representing the isothermal variation of extent of adsorption with pressure as suggested by Freundlich is

$$\left(\frac{x}{m}\right)=kp^{1/n} \qquad \text{(for adsorption of gases)}$$

or

$$\left(\frac{x}{m}\right)=kc^{1/n} \qquad \text{(for adsorption in solution)}$$

where x/m is the mass of adsorbent adsorbed per unit mass of adsorbate and k is constant.

Fundamental Absorption in Vibrational Spectrum

The transition of a diatomic molecule from $\nu=0$ vibrational level to $\nu=1$ vibrational level is known as fundamental absorption.

Fuel Cell

In fuel cells, the cathode and anode constituents are continually supplied and thus electrical energy can be withdrawn indefinitely from a fuel cell as long as the outside supply of fuel is maintained. One of the examples is hydrogen-oxygen fuel cell. The reactions are

Anode $H_2(g)+2OH^-(aq) \longrightarrow 2H_2O(l)+2e^-$
Cathode $O_2(g)+2H_2O(l)+4e^- \longrightarrow 4OH^-(aq)$
Overall $2H_2(g)+O_2(g) \longrightarrow 2H_2O(l)$

One of the advantages of the fuel cells is that the energy is extracted from the reactants under more nearly reversible conditions. This type of cell finds great importance in space vehicles.

G

Galvanic Cell

In the galvanic cell, the conversion of chemical energy into electrical energy takes place. The characteristics of a galvanic cell are as follows:

1. Anode is negative electrode and oxidation reaction takes place at this electrode.
2. Cathode is positive electrode and reduction reaction takes place at this electrode.
3. Current in the external circuit moves from anode to cathode.

Gamma Radiations

The electromagnetic radiations emitted along the emission of α- or β-particles in radioactive decays are referred to as γ-radiations. These rays are highly energetic of very small wavelengths.

Gas Constant *R*

This represents the amount of work (or energy) that can be obtained from one mole of an ideal gas when its temperature is raised by one kelvin. Its value in various units are as follows.

$$R = 8.314 \text{ J K}^{-1} \text{ mol}^{-1}$$
$$= 8.314 \text{ kPa dm}^3 \text{ K}^{-1} \text{ mol}^{-1}$$
$$= 8.314 \text{ MPa cm}^3 \text{ K}^{-1} \text{ mol}^{-1}$$
$$= 8.314 \text{ Pa m}^3 \text{ K}^{-1} \text{ mol}^{-1}$$
$$R = 8.314 \times 10^7 \text{ ergs K}^{-1} \text{ mol}^{-1}$$
$$R = 0.082 \text{ L-atm K}^{-1} \text{ mol}^{-1}$$
$$R = 1.897 \text{ cal K}^{-1} \text{ mol}^{-1}$$

Gas-Ion Half-Cell

In a gas-ion half-cell, an inert collector of electrons, platinum or graphite is in contact with gas and a solution containing a specified ion.

Gibbs Adsorption Isotherm

The Gibbs adsorption isotherm is

$$\Gamma = -\frac{1}{RT} \frac{d\gamma}{d \ln c}$$

where Τ is the excess concentration of solute per unit area of the surface, dγ is the change in the surface tension of the absorbate and c is the concentration of the solution.

Gibbs-Duhem Equation

If the composition of a thermodynamic system is varied, the partial molar quantities of its constituents do not vary independently but in a correlated manner as dictated by Gibbs-Duhem equation. Its expression is

$$\Sigma_i n_i \, dY_{i,pm} = 0$$

where Y may be any partial molar quantity.

Gibbs Function

Gibbs function, G, is defined as

$$G = H - TS$$

This function is also known Gibbs free energy. For a reversible process

$$dG_{T,p} = dw_{net}$$

that is, the decrease of Gibbs free energy for a reversible process at constant T and p is equal to the net amount of non-mechanical work that can be obtained from the system.

For a spontaneous process at constant temperature and pressure, dG of the system follows the expression

$$dG_{T,p} < 0$$

The process continues till the function G attains a minimum value. Thus, at equilibrium $dG_{T,p} = 0$ for any process.

Gibbs-Helmholz Equation

The Gibbs-Helmholtz equation expresses the variation of G or $\triangle G$ with temperature. The equation is

$$G=H+T\left(\frac{\partial G}{\partial T}\right)_p$$

$$\triangle G=\triangle H+T\left(\frac{\partial(\triangle G)}{\partial T}\right)_p$$

It is more convenient to use the variation of $\triangle G/T$ with temperature. The expression of this is

$$\left\{\frac{\partial(\triangle G/T)}{\partial T}\right\}_p=-\frac{\triangle H}{T^2}$$

or

$$\left\{\frac{\partial\,(G\triangle/T)}{\partial\,(1/T)}\right\}_p=\triangle H$$

Glass Electrode

The glass electrode consists of a glass tube terminating in a thin-walled bulb, which contains a solution of constant hydrogen-ion concentration and an electrode of definite potential. Usually, either silver-silver chloride electrode dipped in 0.1 mol L^{-1} solution of hydrochloric acid or a platinum wire inserted in a pH 4.0 buffer solution containing a small quantity of quinhydrone is employed. The bulb is inserted in a solution whose pH is to be determined. The glass electrode potential varies with the pH of solution in which it is immersed and is given by the expression

$$E_{glass}=E^{\circ}_{glass}+\frac{2.303\,RT}{F}\text{pH}$$

The value of E°_{glass} is determined by dipping the glass electrode in the solution of known pH.

Gold Number

Gold number is used to measure the protective power of different colloids. It is defined as the minimum number of milligrams of the dry protective colloid which will check coagulation of 10 cm^3 of a standard red gold sol (i.e. prevent its colour becoming blue) on adding 1 cm^3 of a 10% sodium chloride solution.

Gouy Balance Method

The magnetic susceptibility of a substance can be determined by using Gouy-Balance method. In this method, a tube containing the material is suspended from one arm of a balance in such a way that it is partly in the magnetic field of an electromagnet. When the magnet is turned on, the following two alternatives are observed.

1. If a sample is paramagnetic, it is attracted towards the magnetic field.

2. If a sample is diamagnetic it is repelled by the field.

The force acting on the sample when it is attracted towards the field is given by the expression

$$F=\frac{1}{2}\left(\chi_B - \chi^\circ_B \right) A \frac{B^2_{max}}{\mu_0}$$

where χ_B and χ^0_B are the magnetic susceptibility of the medium and air, respectively, A is the area of cross section of the tube, B_{max} is the maximum value of magnetic flux density and μ_0 is permeability of a vaccum. The force F is measured in terms of the masses that are added or removed from the balance pan in order to keep the position of the sample unshifted, i.e.

$$F=(\triangle m)\, g$$

knowing F, χ_B can be computed from the above expression from which χ_m can be calculated using the expression

$$\chi_m = V_m \chi_B$$

where V_m is the molar volume.

Grahm's Law of Diffusion

The rate of diffusion (or effusion) of a gas is inversely proportional to the square root of its density or molar mass. Mathematically, it is written as

$$\frac{r_1}{r_2} = \sqrt{\frac{\rho_2}{\rho_1}}$$

or
$$\frac{r_1}{r_2} = \sqrt{\frac{M_2}{M_1}}$$

where ρ_1 and ρ_2 represent densities and M_1 and M_2 represent molar masses of the two gases.

Grotthus Conductance

The molar conductivities of H^+ and OH^- in aqueous solution are much larger than those of other ions. This fact is known as Grotthus conductance which has been explained on the basis of jumping of H^+ or OH^- from one water molecule to another. This process of H^+ or OH^- transfer results in a more rapid transfer of charge from one region of the solution to another, than would be possible if the ions H_3O^+ or OH^- has to push its way through the solution as other ions do.

This typ: of mechanism also prevails in any other solvent. Thus, in a given solvent (for example, liquor ammonia), the molar conductivities of its characteristic cation and anion (namely, NH_4^+ and NH_2^-) will have unusally high values than any other cations and anions.

Grotthuss-Draper Law

This law states that "only those radiations which are absorbed by the reacting system are effective in producing chemical change"

It should be clearly understood that though the law states that a photochemical reaction must have resulted because of the absorption of light, the reverse of this is not always true, i.e. the system on absorbing light may or may not result into a chemical reaction. In many cases, the absorbed light is converted into kinetic energy of the absorbing molecules. In many cases, the absorbed light is re-emitted as fluorescence and phosphorescence.

Group Displacement Law

The law may be stated in the following form.

When an α-particle is emitted in a radioactive change, the product is displaced two places to the left in the periodic table, that is, the atomic number is two less than that of its parent, element but the emission of a β-particle results in a displacement of one place to the right, the atomic number of the product being now one greater than that of its parent element.

Group Moments

While dealing with substituted benzenes, the concept of group moment is of great importance. For example, the dipole moment of chlorobenzene is 1.58 D. This is attributed to Cl group, i.e. Cl group moment is 1.58 D. The group moments of other substituents are as follows.

Group	NO_2	CN	Cl	H	CH_3	NH_2	OH
Group moment/D	−3.98	−3.8	−1.58	0	0.04	1.53	1.6

The sign of group moment indicates the direction in which the group moment acts; positive sign implies that the group moment acts away from the benzene ring whereas negative sign implies that it acts towards the ring. The dipole moment of any substituted benzene can be derived from the vector addition of group moments.

Half-Life Time

Half-life time of a reaction is the time required for the concentration of reactant to decrease by half, i.e.

$$\text{at } t=t_{0.5} \qquad [A]_t=[A]_0/2$$

For first order reaction

$$t_{0.5}=\frac{0.693}{k_1}$$

and is independent of initial concentration of reactant.

For second order reaction (2A→B),

$$t_{0.5}=\frac{1}{k_2\,[A]_0}$$

that is $t_{0.5}$ is inversely proportional to the initial concentraiion of A.

Hardy-Schulze Rule

When an electrolyte is added to a colloidal solution, the colloidal particles take up the ions which are oppositely charged and thus they get neutralized. The neutral particles then unite with one another to form bigger particles and thus they get coagulated. Hardy and Schulze observed that the coagulating power of an electrolyte depends upon the valence of ion carrying opposite charge to that of dispersed phase.

Harmonic Oscillator

During the vibration, if the restoring force acting on the atom is given by Hooke's law ($f=-k\,\triangle x$), then the vibration is known as harmonic oscillation. The classical potential energy expression is

$$V=\tfrac{1}{2}k\,(\triangle x)^2$$

where $\triangle x$ is the displacement from the equilibrium position. The quantum mechanical expression is

$$V=(v+\tfrac{1}{2})h\nu$$

where v is the vibrational quantum number (=0, 1, 2,...) and ν, the classical frequency of oscillator, is given by

$$\nu=\frac{1}{2\pi}\sqrt{\frac{k}{\mu}}$$

where μ is the reduced mass of the molecule.

Heat Capacity

Heat capacity of a system is defined as the quantity of heat required to increase its temperature by 1 degree celsius (or kelvin).

For gases, two heat capacities are defined. These are:

Heat capacity at constant volume

and Heat capacity at constant pressure

Helmholtz Function

Helmholtz function, A, is defined as

$$A=U-TS$$

The function A is also called the work function or the work content or the Helmholtz free energy. The name free energy was given to this function by Helmholtz because the decrease in A represents the maximum amount of energy that is free or available for being converted into work, i.e.

$$dA_{T, V}=dw$$

For a spontaneous process at constant temperature and volume, dA of the system follows the expression

$$dA_{T, V}<0$$

The process continues fill the function A attains a minimum value. Thus, at equilibrium $dA_{T, V}=0$ for any process.

Henderson-Hasselbalch Equation

This equation describes the pH of a buffer solution. The equation is

$$\textit{acid buffer}: \mathrm{pH}=\mathrm{p}K_a+\log \frac{[\text{salt}]}{[\text{acid}]}$$

$$\textit{basic buffer}: \mathrm{pOH}=\mathrm{p}K_b+\log \frac{[\text{salt}]}{[\text{base}]}$$

Hermitian Operator

An operator is said to be a hermitian operator if it satisfies the expression

$$\int \psi_n^* \, A_{op} \, \psi_m \, d\tau=\int \psi_m \, A_o{}^*{}_p \, \psi_n^* \, d\tau$$

where ψ_m and ψ_n are the two eign functions of A_{op}. The eigen values of a hermitian operator are real

Hess's Law of Constant Heat Summation

This law states that the heat absorbed or evolved in a given chemical equation is the same whether the process occurs in one step or several steps.

This law is a consequence of first law of thermodynamics. The molar enthalpies of reactants and products involved in a chemical equation have definite values, it is obvious that the enthalpy change of the equation would also have a definite value.

Heterogeneous Catalyst

A heterogeneous catalyst exists in different phase from the reaction it catalyses. Generally, solid surfaces act as heterogeneous catalyst. Examples include decomposition of NH_3 on tungsten, decomposition of N_2O on gold and decomposition of ethanol vapour on the surface of Cu. The rate of a catalytic reaction is usually proportional to the the area of the surface.

Henry's Law

This law states that at a given temperature the mass of dissolved gas in a given volume of solvent is proportional to the pressure of the gas with which it is in equilibrium.

Mathematically, it is written as

$$m \propto p$$

$$m = kp$$

where k is constant of proportionality

For a dilute solution, Henry's law may be stated in terms of mole fraction of gaseous solute in the saturated solution. It states that the solubility of a gas expressed in mole fraction is directly proportional to the pressure of the gas. Mathematically, it is written as

$$x_2 = k'' p_2$$

The above expression is usually written as

$$p_2 = \frac{1}{k''} x_2$$
$$= k_H \, x_2$$

where k_H is known as Henry's law constant.

Homogeneous Catalyst

A homogeneous catalyst exists in the same phase as the reaction whose rate it increases. Examples include acid hydrolysis of an ester, saponification of an ester and enzyme reactions. The rate of a catalytic reaction is usually proportional to the concentration of the catalyst.

Homogeneous Functions

A function $f(x, y, \ldots)$ is said to be a homogeneous function of degree n if the following condition is satisfied.

$$f(\lambda x, \lambda y, \ldots) = \lambda^n f(x, y, \ldots)$$

where λ is an arbitrary parameter and n has a constant integer value.

Hot Band in Vibrational Spectrum

The transitions of a diatomic molecule from $v=1$ vibrational level to higher levels are known as hot bands. These are normally observed at higher temperatures.

Hund's Rule

Hund's multiplicity rule helps us in deciding the relative occupancy of more than one electron in the degenerate orbitals. According to this rule, a configuration with the maximum spin multiplicity has the minimum energy and thus is most stable.

According to this rule, the electrons enter one by one in the degenerate orbitals with the same spin followed by the double occupancy.

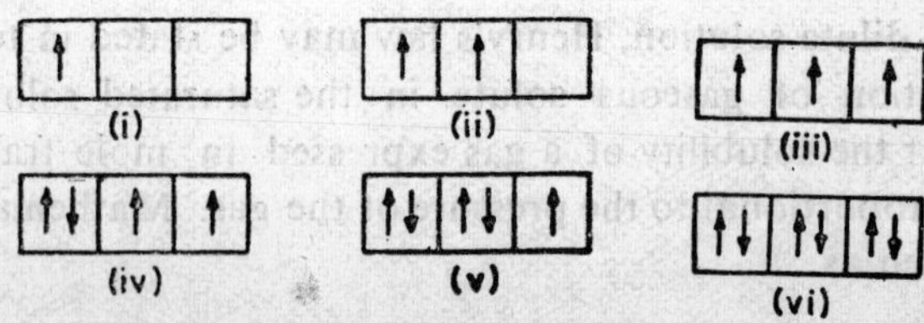

Hund's rule can be understood on the basis of electron-electron interaction between the electrons. By occupying the different orbitals (i.e. quantum number m is different), two electrons are present in different regions and thus the electron-electron repulsive interaction is minimized. By having the same spin, the two electrons have the magnetic moment vectors in the same direction. Thus, the two tiny magnets have like poles near to each other and thereby rebel each other to the maximum extent with the result that the two electrons remain far apart as much as possible. This further minimizes the electron-electron repulsive interaction.

Hybrid Orbitals

The hybrid orbitals are formed by the linear combination of atomic orbitals centred on the same atom. The number of hybrid orbitals formed are equal to the number of atomic orbitals being mixed. The formed hybrid orbitals are completely identical is size, shape and energy. Their orientation around the nucleus is completely symmetrical. These orbitals are better than atomic orbitals as far as bonding with other orbital is concerned.

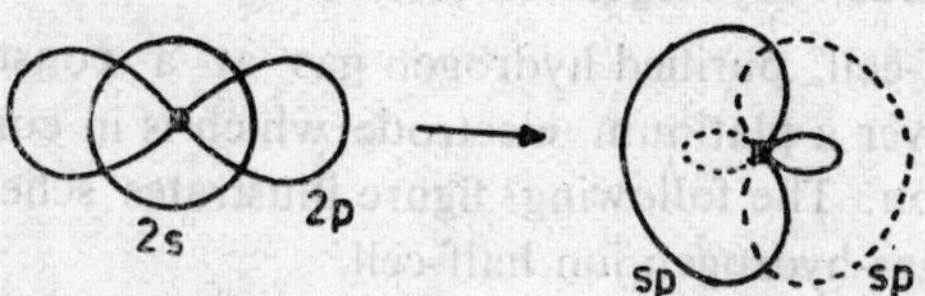

The geometrical arrangement of atoms in the molecule is explained on the basis of hybrid orbitals. For example, CH_4 involves tetrahedral geometry because sp^3 hybrid orbitals of C are involved in the bonding with hydrogen atoms. Ethylene molecule is planar because sp^2 hybrid orbitals of C are involved in the bonding. The following is the brief description of type of hybrid orbital involved in the bonding and the resultant molecular geometry.

Hybridization	*shape*	*Examples*
sp	linear	acetylene, BeH_2
sp^2	planar	BCl_3, ethylene
sp^3	tetrahedral	methane
dsp^2	square planar	$Ni(CN)_4^{2-}$
d^2sp^3	bipyramidal	PF_5
d^2sp^3	octahedral	SF_6

Hydrogen Bond

In a molecule, if hydrogen is covalently attached to a highly electronegative atom such as fluorine, oxygen or nitrogen, the bond is polarized and thus hydrogen acquires a small positive charge, whereas the electronegative atom acquires a small negative charge. The slightly positively charged hydrogen atom gets attached to an adjacent highly electronegative atom (containing lone pair) through a weak directional bond, known as the hydrogen bond. This may be represented as

$$\overset{\delta-}{A}—\overset{\delta+}{H}\ldots\overset{\delta-}{A}—\overset{\delta+}{H}\ldots$$

Hydrogen Gas—Hydrogen Ion Half-Cell

In this half-cell, purified hydrogen gas at a constant pressure is passed over a platinum electrode which is in contact with an acid solution. The following figure illustrates schematically the hydrogen gas-hydrogen ion half-cell.

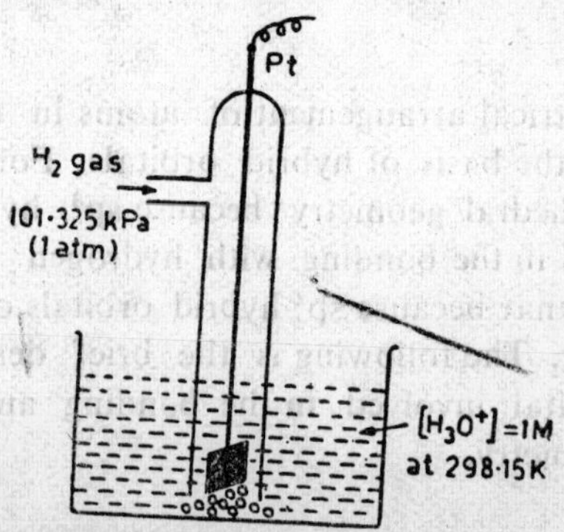

Hydrogen Spectrum

Balmer observed that the wavelengths of emitted radiations in the visible and near-ultraviolet region of the hydrogen spectrum could be expressed as

$$\lambda=(364.56 \text{ nm})\left(\frac{n_2^2}{n_2^2-n_1^2}\right)$$

where $n_1=2$ and $n_2=3, 4, 5, \ldots$. In the wave number units, the expression is

$$\nu=R_H\left(\frac{1}{n_1^2}-\frac{1}{n_2^2}\right)$$

where R_H is known as Rydberg constant and it has a value of 1.09737×10^5 cm^{-1}.

Five spectral series of hydrogen atoms are known. These are as follows.

Lyman series	ultra violet spectral region	$n_1=1$	$n_2=2, 3, 4, \ldots$
Balmer series	visible region	$n_1=2$	$n_2=3, 4, 5, \ldots$
Paschen series	near infrared region	$n_1=3$	$n_2=4, 5, 6, \ldots$
Brackett series	infrared region	$n_1=4$	$n_2=5, 6, 7, \ldots$
Pfund series	far infrared region	$n_1=5$	$n_2=6, 7, \ldots$

Hydrolysis of Anions

The anion A^- which is a weaker base than OH^- and which has its conjugate acid HA stronger than water but weaker than H_3O^+ undergoes hydrolysis according to the equation

$$A^-+H_2O\rightleftharpoons HA+OH^-$$

The above equilibrium is characterized by a constant, known as hydrolysis constant. It is defined by the expression

$$K_h=\frac{[HA]\,[OH^-]}{[A^-]}$$

$$=\frac{K_w}{K_a}$$

where $K_w=[H^+]\,[OH^-]$

and $K_a=\frac{[H^+]\,[A^-]}{[HA]}$

Hydrolysis of Both Cations and Anions

A salt formed from a weak acid and a weak base (such as ammonium acetate) is an example in which both its cation and anion undrgo hydrolysis:

$$B^+ + A^- + H_2O \rightleftharpoons BOH + HA$$

Its hydrolysis constant is given by the expression

$$K_h=\frac{[BOH]\,[HA]}{[B^+]\,[A^-]}$$

$$=\frac{K_w}{K_aK_b}$$

where $K_w=[H^+]\,[OH^-]$

$K_a=[H^+]\,[A^-]/[HA]$

and $K_b=[B^+]\,[OH^-]/[BOH]$

Hydrolysis of Cations

The cation B^+ which is a weaker acid than H_3O^+ and which has its conjugate base BOH stronger than water but weaker than OH^- undergoes hydrolysis according to the equation

$$B^+ + H_2O \rightleftharpoons BOH + H^+$$

The above equilibrium is characterized by a constant, known as hydrolysis constant. It is defined by the expression

$$K_h=\frac{[BOH]\,[H^+]}{[B^+]}$$

$$=\frac{K_w}{K_b}$$

where $K_w=[H^+]\,[OH]$

and $K_b=\dfrac{[B^+]\,[OH^-]}{[BOH]}$

Hydrolysis of Multivalent Cation

Hydrolysis of multivalent cations takes place in a stepwise manner and more than one hydrolytic products are formed. For example, for Fe^{2+} ions we have

$$Fe^{2+}+2H_2O \rightleftharpoons Fe(OH)^+ + H_3O^+$$
$$Fe(OH)^+ + 2H_2O \rightleftharpoons Fe(OH)_2 + H_3O^+$$

Normally, first hydrolysis is much stronger than the second and hence $K_{h1} >> K_{h2}$.

Hydrolysis of Multivalent Anion

Hydrolysis of multivalent anions takes place in a stepwise manner and more than one hydrolytic products are formed. For example, for S^{2-} ions we have

$$S^{2-}+H_2O \rightleftharpoons HS^- + OH^-$$
$$HS^- + H_2O \rightleftharpoons H_2S + OH^-$$

Normally, first hydrolysis is much stronger than the second and hence $K_{h1} >> K_{h2}$.

Hydrolysis of Salts

A given salt on dissolving in water may produce acidic, neutral or alkaline solution depending upon its ions. This is due to the fact that certain ions can react with water and thereby produce an acidic or an alkaline solution according to the following reactions.

$$A^- + H_2O \rightleftharpoons HA + OH^-$$
$$B^+ + 2H_2O \rightleftharpoons BOH + H_3O^+$$

This phenomenon is known as hydrolysis.

Hyperfine Structures in ESR Spectrum

The interaction of electron with the magnetic active nuclei produces hyperfine structures in ESR spectrum, in which more than one absorption is observed. Interactions with the equivalent protons producs $n+1$ absorptions, where n is the number of equivalent protons. Their relative intensities are given by the expression

$$C^n{}_m=\frac{n!}{m!(n-m)!}$$

where m takes the values of 0, 1, 2,..., n.

Hypsochromic Shift

The general phenomenon of the shift of an absorption band to a region of shorter wave length is often referred to as hypsochromic shift.

I

Ideal Gas

An imaginary gas to which all experimentally derived gas laws (Boyle's, Charles, etc.) are applicable under all conditions of temperature and pressure. A real gas behaves more or less ideally under the condition of low pressure and high temperature.

Ideal Law of Solubility

The expression of ideal law of solubility is

$$\ln x_2 = -\frac{\triangle H_{m,fus}}{R}\left[\frac{1}{T}-\frac{1}{T_f{}^*}\right]$$

where $\triangle H_{m,fus}$ is the molar enthalpy of fusion of solute, $T_f{}^*$ is the freezing point of pure solute. The ideal solubility is independent of the nature of solvent.

Ideally Dilute Solution

A real solution in the limit $x_{solvent} \rightarrow 1$ and $x_{solute} \rightarrow 0$ is known as ideally dilute solution. Raoult's law for solvent and Henry's law for solute are the defining equations for an ideally dilute solution, i.e.

Raoult's law for solvent: $p_1 = x_1 p_1^*$

Henry's law for solute: $p_2 = x_2 K_H$

where p_1^* is the vapour pressure of pure solvent and K_H is the Henry's law constant.

Ideal Solutions of Liquids in Liquids

A solution is said to be an ideal solution if its constituents follow Raoult's law under all conditions of composition, i.e. partial pressure of each and every constituent is given by

$$p_i = x_i \, p_i^*$$

where p_i is the partial pressure of the constituent i whose mole fraction in the solution is x_i and p_i^* is the corresponding vapour pressure of the pure constituent.

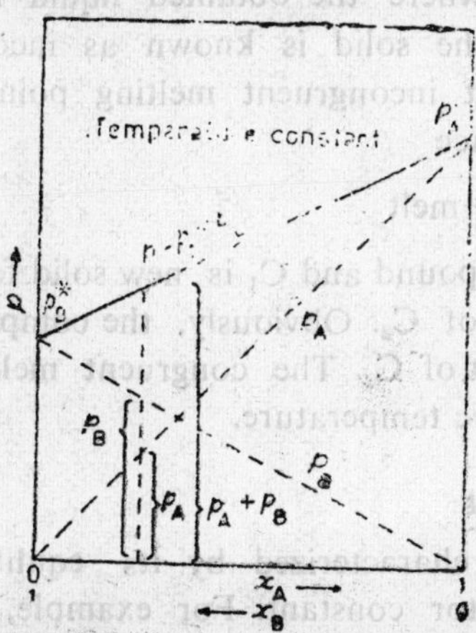

Ionic Product and Precipitation

For a solution of a salt at a specified concentration, the product of concentration of its ions, each raised to a power equal to the corresponding stoichiometric number appeared in the chemical equation, is known as ionic product. For a saturated solution in equilibrium with excess solid, the ionic product is equal to its solubility product. If the ionic product of the solution is less than the corresponding solubility product, then the solution is unsaturated and thus more salt can be dissolved in it. On the other hand, if the ionic product exceeds the solubility product, then the solution is holding more salt that can dissolve in it; therefore precipitation takes place which continues till the ionic product becomes equal to the solubility product. In brief, we have

ionic product > solubility product ⇒ precipitation

ionic product = solubility product ⇒ saturated solution

ionic product < solubility product ⇒ unsaturated solution

Incongruently Saturating Double Salt

If a double salt is no longer stable in the presence of water, it is said to be an incongruently saturating double salt.

Incongruent Melting Point

A melting point where the obtained liquid has composition different from the solid is known as incongruent melting point. In fact, at incongruent melting point, the following equation takes place

$$C_2 \rightleftharpoons C_1 + \text{melt}$$

where C_2 is a compound and C_1 is new solid formed as a result of decomposition of C_2. Obviously, the composition of melt is different from that of C_2. The congruent melting point is also known as peritectic temperature.

Indicator Constant

An indicator is characterized by its equlibrium constant, known as indicator constant. For example, for an organic weak acid, we have

$$HIn \rightleftharpoons H^+ + In^-$$

$$K_{In} = \frac{[H^+]\,[In^-]}{[HIn]}$$

Indicator Range

If the solution contains two coloured species, say In^- and HIn, the colour of the solution depends on the relative amounts of the two species. On an average, the solution acquires a distinct colour characteristic of In^- if the concentration of the latter is approximately 10 or more times greater than that of HIn and vice versa.

The relative ratio of $[In^-]$ and [HIn] in a solution is pH dependent and is given by the expression

$$pH = pK_{I_n} + \log \frac{[In^-]}{[HIn]}$$

For $pH \geqslant pK_{I_n}+1$, the ratio $\frac{[In^-]}{[HIn]} \geqslant \frac{10}{1}$. Hence, the solution gets the colour characteristic of In^-.

For $pH \leqslant pK_{I_n}-1$, the ratio $\frac{[In^-]}{[HIn]} \leqslant \frac{1}{10}$. Hence, the solution get the colour characteristic of HIn.

The range of about $pK_{I_n}-1$ to $pK_{I_n}+1$ is known as indicator range. For example, for phenolphthalene indicator range is 8.3 (colourless) to 10.0 (red), methyl orange is 3.1 (red) to 4.4 (yellow) and methyl red is 4.2 (red) to 6.3 (yellow).

Inexact Differential

In the differential expression

$$d\phi = P_{x,y}\, dx + Q_{x,y}\, dy$$

If $\left(\frac{\partial Q}{\partial x}\right)_y \neq \left(\frac{\partial P}{\partial y}\right)_x$

then the differential $d\phi$ is known as inexact differential.

Instability Constant

The dissociation of complex takes place in a step wise manner. The product of these constants for a particular complex gives an overall constant which is known as instability constant. For example, for $Ag(NH_3)_2^+$ we have

$$Ag(NH_3)_2^+ \rightleftharpoons Ag(NH_3)^+ + NH_3$$

$$K_1 = \frac{[Ag(NH_3)^+]\,[NH_3]}{[Ag(NH_3)_2^+]}$$

and $\quad Ag(NH_3)^+ \rightleftharpoons Ag^+ + NH_3$

$$K_2 = \frac{[Ag^+]\,[NH_3]}{[Ag(NH_3)^+]}$$

Hence,

$$K_{inst} = K_1 K_2 = \frac{[Ag^+]\,[NH_3]^2}{[Ag(NH_3)_2^+]}$$

which corresponds to the overall reaction

$$Ag(NH_3)_2^+ \rightleftharpoons Ag^+ + 2NH_3$$

Instantaneous Rate of Change of the Amount of a Reactant or a Product

In chemical kinetics, the rate at any particular instant rather than the average rate over a time interval has much more practical application and importance. The rate is known as the instantaneous rate (or simply rate) and is defined as

$$R_{ins} = \lim_{\Delta t \to 0} \left(-\frac{\Delta n_A}{\Delta t} \right)_t = -\frac{dn_A}{dt}$$

$$R'_{ins} = \lim_{\Delta t \to 0} \left(\frac{\Delta n_B}{\Delta t} \right)_t = \frac{dn_B}{dt}$$

where A and B are reactant and product, respectively.

Integral Enthalpy of Dulution

The integral enthalpy of dilution is the change in enthalpy when a solution containing 1 mole of solute is diluted from one concentration to another. According to Hess's law, it is equal to the difference between the integral enthalpies of solution at two concentrations.

Integral Enthalpy of Solution

The integral enthalpy of solution at a given concentration is defined as the enthalpy change when one mole of the solute is dissolved in a definite quantity of solvent to produce a solution of a particular concentration.

Integrating Factor

An inexact differential may be converted into exact differential with the help of integrating factor, For example, an inexact differential

$$d\phi = y\,dx - x\,dy$$

becomes exact differential if the above equation is multiplied by

$1/x^2$ [or $1/y^2$ or $1/xy$ or $1/(x^2+y^2)$]

throughout, i.e. in the expression

$$d\phi' = \frac{1}{x^2}\,d\phi = \frac{y}{x^2}\,dx - \frac{1}{x}\,dy$$

$d\phi'$ is an exact differential and $1/x^2$ is the integrating factor.

Integrated Rate Laws

The integrated rate laws express how exactly the concentration of a species changes with time. These laws are obtained by integrating differential rate laws. A few Integrated rate laws for different order of a reaction are given below.

Order	*Integrated rate law*
zero	$[A]_t = [A]_o - x$
first	$\ln [A]_t = \ln [A]_o - kt$
second	$\frac{1}{[A]_t} = \frac{1}{[A]_o} + k_2t$
	or $\log \frac{[B]_t}{[A]} = \log \frac{[B]_o}{[A]} + \frac{[B]_o - [A]_o}{2.303} k_2t$

Intensive Variable

The variables which are independent of mass of the system are known as intensive variables. Examples are; temperature, pressure, concentration, density, dipole moment, refractive index, viscosity, surface tension, molar volume, gas constant, specific heat capacity, vapour pressure, dielectric constant and emf of a cell.

Interpretation of $\triangle U$ in Terms of Heat Exchange at Constant Volume

From the first law of thermodynamics, we have

$$\mathrm{d}U = \mathrm{d}q + \mathrm{d}w$$

Now $\mathrm{d}w = -p_{ext}\mathrm{d}V$

Hence $\mathrm{d}U = \mathrm{d}q - p_{ext}\,\mathrm{d}V$

For a process at constant volume, $-p_{ext}\mathrm{d}V$ is equal to zero. Hence

$$\mathrm{d}U = \mathrm{d}q_V$$

or $\triangle U = q_V$

that is, heat transferred at constant volume by a system changes its internal energy.

Inversion Temperature

The temperature at which $\mu_{JT} = 0$ in Joule-Thomson experiment is known as inversion temperature. For a van der waals gas, the inversion temperature is given by the expression

$$Ti = \frac{2a}{Rb}$$

where a and b are van der waals constants and R is the gas constant

Ionic bond

A bond formed by the transfer of one electron(s) from electropositive atom to the electronegative atom is said to be an ionic bond. In the above process, ions are formed and these are held together by electrostatic interactions.

Ionic Mobility

The velocity of ion under the influence of unit potential gradient is known as ionic mobility. The ionic mobility of an ion is given by the expression

$$u = \frac{\lambda}{zF}$$

where λ is the ionic molar conductivity of the ion, z is its charge number and F is Faraday constant (=96487 C mol^{-1}).

Ionic Product of Water

The degree of dissociation of water at 25 °C has a very small value of 1.8×10^{-9}. For such a small ionization, the concentration of unionized water remains practically the same. Its value is

$$c = \frac{1000 \text{ g dm}^{-3}}{18 \text{ g mol}^{-1}} = 55.56 \text{ mol dm}^{-3}$$

This concentration when combined with the equlibrium constant of water gives a new constant known as ionic product of water. It is defined as

$$K_w = [H^+]\,[OH^-]$$

Its value at 25 °C is given below.

$$K_w = K_{eq}\,[H_2O] = \left(\frac{1}{55.56}\times10^{-14} \text{ mol dm}^{-3}\right)\times(55.56 \text{ mol dm}^{-3})$$

$$= 1.0\times10^{-14}\,(\text{mol dm}^{-3})^2$$

Ionic Strength of Solution

The ionic strength of the solution is defined as

$$\mu = \tfrac{1}{2}\Sigma_i\, c_i\, z_i^2$$

c_i is the concentration of the ith ions in mol dm^{-3} and z_i is its charge number. The summation is to be carried over all types of ions present in the solution.

Ionization Constant of Water

Pure water itself is a weak electrolyte and ionizes according to the equation

$$H_2O + H_2O \rightleftharpoons H_3O^+ + OH^-$$

or simply written as

$$H_2O \rightleftharpoons H^+ + OH^-$$

This equilibrium is characterized by its equilibrium constant defined as

$$K_{eq} = \frac{[H^+]\,[OH^-]}{[H_2O]}$$

Its value at 25 °C is $(1/55.56) \times 10^{-14}$ mol dm^{-3}.

Isobars

Atoms having the same mass number but different atomic number are known as isobars.

Ionization Constant of a Weak Acid

See dissociation constant of a weak acid.

Ionization constant of a Weak Base

See dissociation constant of a weak base

Isoelectric Point

An ampholyte is at its isoelectric point when the concentration of positive ions RH^+_2 is equal to that of the negative ions $R\gamma^-$. At the isoelectric point, an ampholyte appears to remain stationary in an electric field.

Isomorphism

The isomorphism is a phenomenon in which one element can be replaced by another of a related chemical type without producing any appreciable change in crystal form. Examples are $KH_2PO_4.H_2O$. and $KH_2AsO_4.H_2$.

Isopleth

Any vertical line in a temperature-composition phase diagram is known as isopleth (From Greek for same abundance) Along the isopleth, the temperature of a system having overall constant composition varies.

Isothermal Process

A process in which temperature of the system remains the same is known as isothermal process

Isotopes

The atoms having the same atomic number but different mass numbers are known as isotopes.

Joule-Thomson Coefficient

The Joule-Thomson coefficient μ_{JT} is defined as

$$\mu_{JT} = \left(\frac{\partial T}{\partial pH}\right)$$

$$\mu_{JT} = \frac{V}{C_p}\left[\frac{T}{V}\left(\frac{\partial V}{\partial T}\right)_p - 1\right]$$

For an ideal gas

$$\mu_{JT} = 0$$

Joule-Thomson Experiment

The Joule-Thomson experiment involves the expansion of a gas from one fixed pressure to another under adiabatic conditions. This expansion is an isenthalpic (i.e. constant enthalpy) process.

Kelvin

The kelvin, unit of thermodynamic temperature, is the fraction 1/273.16 of the thermodynamic temperature of the triple point of water.

Kelvin-Planck Statement of Second Law of Thermodynamics

The Kelvin-Planck statement of second law of thermodynamics is as follow.

It is impossible for a system operating in a cycle and connected to a single heat reservoir to produce a positive amount of work in the surroundings.

Kelvin Temperature Scale

Kelvin showed that it is possible to introduce a thermodynamic temperature scale which is independent of the material used for the thermometric substance. Base on carnot cycle, the temperature scale is given by the expression

$$\theta_2=(273.16\text{ K})\left(\frac{q_2}{273.16}\right)$$

where 273.16 K is the temperature assigned to the triple point of water and q_2 and $q_{273\cdot 16}$ are the heats involved in the isothermal processes of carnot cycle at temperature θ_2 and 273.16 K, respectively. Kelvin temperature is found to be identical with that established on the basis of ideal gas. The kelvin temperature is related to Celsius temperature through the expression

$$T/\text{K}=t/^{\circ}\text{C}+273.15$$

Kilogram

The kilogram is the unit of mass, it is equal to the mass of the international prototype of the kilogram.

Kinetic Gas Equation

This is a theoretical expression deriveable from the characteristics of an ideal gas. Its expression is

$$pV=\tfrac{1}{3}mN\,\overline{u^2}$$

where m is the mass of a single molecule of an ideal gas, N is the number of molecules in the gaseous system and $\overline{u^2}$ is the mean square speed of gaseous molecules.

K irchoff's Relation

T he enthalpy of a reaction at a temperature different from hat at which the value is available can be computed from

Kirchoff's relation. Its expression is

$$\triangle H=\triangle H_0+\int(\triangle C_p)dT$$

where $\triangle H_0$ is the constant of integration and $\triangle C_p$ is the change in heat capacity involved in a chemical equation. For example, for the equation

$$aA+bB+...=lL+mM+...$$

the expression of

$$\triangle H=(lH_{m,L}+mH_{m,M}+...)-(aH_{m,A}+bH_{m,M}+...)$$

Kohlrausch's Law

This law is also known as Kohlrausch's law of independent migration of ions. This law states that at infinite dilution, where dissociation for all electrolytes is complete (including the weak electrolytes since degree of dissociation approaches one as the concentration approaches zero, Ostwald dilution law) and where all interionic effects disappear (because of large distance between ions), each ion migrates independently of its co-ion and contributes to the total molar conductivity of an electrolyte a definite share which depends on its own nature and not at all on ion with which it is associated. Thus, the molar conductivity at infinite dilution is written as

$$\overset{\infty}{\Lambda}_m=\nu_+\overset{\infty}{\lambda}_++\nu_-\overset{\infty}{\lambda}_-$$

where ν_+ and ν_- are the stoichiometric numbers of cation and anion associated with a given electrolyte and $\overset{\infty}{\lambda}_+$ and $\overset{\infty}{\lambda}_-$ are the corresponding molar ionic conductivities.

Konowaloff's Rule

A general conclusion regarding the composition of vapour phase relative to that of the liquid phase can be derived thermodynamically. However, this rule was derived empirically by D.P. Konowaloff on the basis of his systematical measurement of total vapour pressure of homogeneous liquid systems. The rule is as follows.

The vapour phase is richer in the component whose addition to the liquid mixture results in an increase of total vapour pressure or alternatively, the liquid phase is richer in the component whose addition to the liquid phase results in a decrease of total vapour pressure.

Since the boiling point of a solution is inversely related to its vapour pressure (higher vapour pressure means lower boiling point and vice versa), the Konowaloff's rule can be stated as follows.

The vapour phase is richer in the component whose addition to the liquid mixture causes a decrease in its boiling point, or alternatively, the liquid phase is richer in the component whose addition to the liquid mixture causes increase in its boiling point.

L

Lambert-Beer's Law

The amount of light absorbed by a solution is

$$I = I_0^{-\varepsilon lc}$$

where I is intensity of transmitted light and I_0 is that of incident light, ε is the molar absorption coefficient (formely as molar extinction coefficient or molar absorptivity), l is the length of the photochemical cell and c is the concentration of light absorbing solution.

Lambert Law

The amount of light absorbed by a pure substance follows Lambert law, according to which, we have

Equal fractions of the incident radiation are absorbed by successive layers of equal thickness of the light absorbing substance.

Mathematically, it is written as

$$-\frac{dI}{I} \propto dl$$

or $$-\frac{dI}{I} = k\ dl$$

On integrating this expression, we get

$$\ln\left(\frac{I}{I_0}\right) = -kl$$

or $$I = I_0 e^{-kl}$$

or $$I = I_0\ 10^{-al}$$

where $a = k/2.303$ and is known as absorption coefficient (formerly as extinction coefficient or absorptivity).

Laminar Flow

If a fluid in a tube is flowing layerwise, the flow is said to be a laminar flow.

Langmuir Adsorption Equation

Langmuir adsorption equation describes the variation in the extent of adsorption with pressure (or concentration) of adsorbent. The equation is

$$\frac{x}{m} = \frac{K_1 k_2 p}{1 + K_1 p}$$

where x/m is the mass of adsorbent adsorbed per unit mass of adsorbate and k_1 and k_2 are constants.

Law of Corresponding States

According to this law, if two gases have the same values of reduced pressure and reduced temperature, they will have the same reduced volume. This law is contained in a general expression, known as reduced equation of state, which is applicable to all gases;

$$\left(p_r + \frac{3}{V_r}\right)(3V_r - 1) = 8T_r$$

where $p_r = p/p_c$

$V_r = V_m/V_c$

$T_r = T/T_c$

where p_c, V_c and T_c are the critical pressure. critical volume ahd critical temperature of the gas.

Law of Distribution of Molecular Speeds

This law expresses the manner in which the molecules of a gas are distributed over the possible speed ranges, from zero to very high values. This law was investigated by J.C. Maxwell. The expression of this law is

$$dN = 4\pi N\left(\frac{M}{2\pi RT}\right)^{3/2} \exp(-Mu^2/2RT)u^2 du$$

where the various symbols have their usual meanings. The following figure illustrates the typical distribution of speeds.

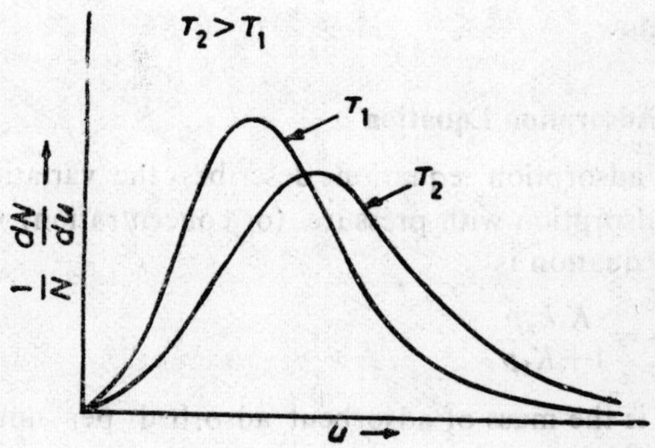

Law of Equipartition of Energy

The average energy of a molecule can be calculated with the help of the classical law of equipartition of energy, according to which, we have

If the energy of a molecule can be written in the form of a sum of terms, each of which is proportional to the square of a velocity component (or to the square of a position coordinate), then each of these square terms contribute $(\frac{1}{2})kT$ to the average energy.

According to this law, we have

$$E_{\text{translation}} = \tfrac{1}{2}mu^2_x + \tfrac{1}{2}mu^2_y + \tfrac{1}{2}mu^2_z$$
$$= \tfrac{1}{2}kT + \tfrac{1}{2}kT + \tfrac{1}{2}kT$$
$$= \frac{3}{2}kT$$

$$E_{\text{rotation}} = \tfrac{1}{2}I\omega^2_x + \tfrac{1}{2}I\omega^2_y$$
(linear molecule)
$$= \tfrac{1}{2}kT + \tfrac{1}{2}kT$$
$$= kT$$

$$E_{\text{rotation}} = \tfrac{1}{2}I\omega^2_x + \tfrac{1}{2}I\omega^2_y + \tfrac{1}{2}I\omega^2_z$$
(nonlinear molecule)
$$= \tfrac{1}{2}kT + \tfrac{1}{2}kT + \tfrac{1}{2}kT$$
$$= \frac{3}{2}kT$$

$$E_{\text{vibration}} = \text{(kinetic + potential) energy}$$
$$= \tfrac{1}{2}\mu\,(dr/dt)^2 + \tfrac{1}{2}k(r - r_0)^2$$
$$= \tfrac{1}{2}kT + \tfrac{1}{2}kT$$
$$= kT$$

Law of Mass Action

According to the law of mass action, the rate of a reaction is proportional to the product of effective concentrations of the reacting species, each raised to a power which is equal to the corresponding stoichiometric number of the substance appearing in the chemical equation

For example, for the chemical equation

$$\nu_1A_1 + \nu_2A_2 \rightleftharpoons \nu_3A_3 + \nu_4A_4$$

The rate of forward reaction is

$$r_f = k_f\,[A_1]^{\nu_1}\,[A_2]^{\nu_2}$$

The rate of backward reaction is

$$r_b = k_b[A_3]^{\nu_3}\,[A_4]^{\nu_4}$$

At equilibrium

$$r_f = r_b$$

Hence

$$K_{eq}=\frac{k_f}{k_b}=\frac{[A_3]^{\nu 3}\,[A_4]^{\nu 4}}{[A_1]^{\nu 1}\,[A_2]^{\nu 2}}$$

Law of Photochemical Equivalence

The law was proposed by Stark and Einstein. According to this law, we have

Each light absorbing molecule in a photo-chemical reaction absorbs only one quantum of light causing the activation.

Law of Rational Indices

To describe the orientation of planes passing through lattice points of a crystal, three basic unit lengths a, b and c are chosen along the three basis directions. It is found that the ratio of intercepts h' k' and l' of a plane on the three basis directions and the corresponding unit length is either an integer or a ratio of two integers. This statement is known as law of rational indices.

LCAO

One of the methods to construct molecular orbitals of a molecule is to take the linear combination of atomic orbitals (abbreviated as LCAO). For example, for hydrogen molecule, we can write

$$\psi_{mo}=C_1\psi_{1s}(H_a)+C_2\psi_{1s}(H_b)$$

where $\psi_{1s}(H_a)$ and $\psi_{1s}(H_b)$ are the wave function of 1s orbitals centred on atom H_a and H_b, respectively.

Lead Storage Cell

In this cell, lead acts as anode and lead impregnated with lead dioxide acts as cathode. The electrolyte is a solution of approximately 20 per cent sulphuric acid with a specific gravity of about 1.15 at room temperature. The reactions involved are

Anode $Pb(s)+SO_4^{2-}(aq)\longrightarrow PbSO_4(s)+2e^-$

Cathode $PbO_2(s)+4H^++SO_4^{2-}(aq)+2e^-$
$\longrightarrow PbSO_4(s)+2H_2O(l)$

Overall $Pb(s) + PbO_2(s) + 2H_2SO_4$
$\longrightarrow 2PbSO_4(s) + 2H_2O(l)$

The emf of the cell depends on the concentration of sulphuric acid in solution. At 25 °C, some of the values are

1.90 V at 7.4% H_2SO_4

2.0 V at 21.4% H_2SO_4

2.14 V at 39.2% H_2SO_4

Le Chatelier Principle

See, principle of Le Chatelier and Braun

Lever Rule

The lever rule gives the relative amounts of two phases in equilibrium with each other. For example, for the system shown in the following Figure, the relative amounts of liquid and vapour phases are as follows.

$$\frac{\text{Amount of vapour phase}}{\text{Amount of liquid phase}} = \frac{al}{av}$$

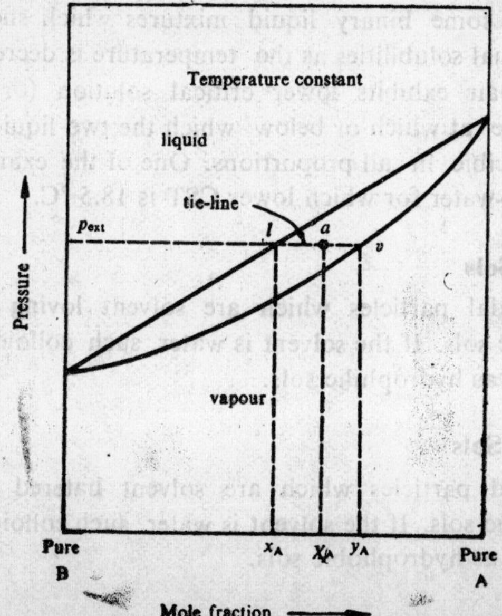

Liquid Junction Potential

In a cell if two electrolytic solutions of different concentrations are in contact with each other, a potential difference develops across the boundary of the two solutions. This potential difference is called the liquid function potential or the diffusion potential.

Lorentz-Lorenz Equation

Lorentz-Lorenz equation is

$$\frac{n^2-1}{n^2+2}\frac{M}{\rho}=\frac{N_A}{3\varepsilon_0}\alpha_e$$

where n is the refractive index of the medium, ε_0 is the permittivity of a vacuum, M/ρ is the molar volume of the medium, N_A is Avogadro constant and α_e is the electronic polarizability of the molecule. The left side of the above expression is known as molar refraction and is represented by the symbol R_m.

Lower Consolute Temperature

There are some binary liquid mixtures which show increase in the mutual solubilities as the temperature is decreased. Such a liquid pair exhibits lower critical solution (or consolute) temperature at which or below which the two liquids are completely miscible in all proportions. One of the examples is triethylamine-water for which lower CST is 18.5 °C.

Lyophilic Sols

The colloidal particles which are solvent loving are known as lyophilic sols. If the solvent is water, such colloidal particles are known as hydrophilic sols.

Lyophobic Sols

The colloid particles which are solvent hatered are known as lyophobic sols. If the solvent is water, such colloidal particles are known as hydrophobic sols.

M

Magnetic Moment of a Nucleus

The magnetic moment of a nucleus which is due to its spinning is given by the expression

$$\mu_m = g\,\mu_N \sqrt{I(I+1)}$$

where *g* is known as nuclear splitting factor, μ_N is nuclear magneton ($= eh/4\pi m_p$) and *I* is spin quantum number.

For nuclei having $I=0$ (e.g. $^{12}_{6}C$ and) $^{16}_{8}O$, μ_m is zero. Such nuclei are known as magnetic inactive nuclei.

Magnetic Permeability

In a vacuum the magnetic flux density *B* is related to the magnetic field intensity *H* by the relation

$$B = \mu_0 H$$

where μ_0 is known as magnetic permeability of the vacuum.

In SI units, μ_0 has the value of $4\pi \times 10^{-7}$ N C^{-2} s^2.

Magnetic Quantum Number

This number (symbol *m*) describes the *z*-component of the angular momentum of the electron through the expression

$$L_z = m\left(\frac{h}{2\pi}\right)$$

This number dictates the direction of orbital with respect to *z*-axis. The permitted values of *m* are

$$-l, -l+1, \ldots, l-1, l$$

The *z*-axis, by convention, is the direction of electrical or magnetic field.

The energy of electron in hydrogen-like atoms depends only on the value of *n*. All orbitals with the same value of *n* but different values of *l* and *m* are degenerate. For a multi-electron atom, energy depends on both the quantum numbers *n* and *l*.

Magnetic Susceptibility

The magnetization of a substance is directly proportional to the applied magnetic field and is given by the expression

$$M=\chi_B\ (B/\mu_0)$$

where χ_B is known as magnetic susceptibility of the substance.

Maxwell's Law of the Distribution of Molecular Speeds

Maxwell deduced a theoretical expression of molecular speeds based on statistical behaviour of molecules. The expression is

$$\frac{dN_c}{N}=4\pi\left(\frac{m}{2\pi kT}\right)^{3/2}\exp(-mc^2/2kT)c^2dc$$

where dN_c/N is the fraction of molecules having speed between c and $c+dc$, m is the mass of each molecule and k is Boltzmann constant.

Maxwell Relations

The maxwell relations equate the rate of change of a particular quantity (which cannot be determined experimentally) with those experimentally observable quantities of pressure, volume and temperature. The relations are

$$\left(\frac{\partial S}{\partial p}\right)_V=-\left(\frac{\partial V}{\partial T}\right)_S$$

$$\left(\frac{\partial S}{\partial V}\right)_p=\left(\frac{\partial p}{\partial T}\right)_S$$

$$\left(\frac{\partial S}{\partial V}\right)_T=\left(\frac{\partial p}{\partial T}\right)_V$$

$$\left(\frac{\partial S}{\partial p}\right)_T=-\left(\frac{\partial V}{\partial T}\right)_p$$

Mean Free Path

Mean free path is defined as the average distance a molecule moves between the two successive collisions. Its expression is

$$\lambda=\frac{1}{\sqrt{2}\pi N\sigma^2}$$

where N is the number of molecules per unit volume of the gas and σ is the collision diameter.

Metal-Insoluble Salt-Anion Half-Cell

In this half-cell, a metal coated with its insoluble salt is in contact with a solution containing the anion of the insoluble salt. Three such half-cells are commonly used.

1. Silver-silver chloride-chloride ion half-cell
2. Mercury-mercuric oxide-hydroxide ion half cell
3. Mercury-mercurous chloride-chloride ion half-cell. This cell is also known as calomel cell.

Metal-Metal Ion Half-Cell

Metal-metal ion half-cell consists of a bar of metal M in contact with a solution containing M^{n+} ions. Examples include zinc-zinc ion, copper-cupric ion, silver-silver ion and gold-auric ion half-cells.

Metastable Phase

A substance which exists in a phase which is not normally a stable phase at that temperature and pressure, the phase is said to be metastable phase. For example, at -1 °C and 1 atm, the stable phase of water is the solid phase (ice). However, water may exist in the form of liquid phase at these conditions. Thus, liquid phase of water at -1 °C and 1 atm is said to be a metastable phase. The metastable phase generally is unstable and it passes over to the stable phase by slight disturbance. Its vapour pressure is larger than the corresponding vapour pressure of the stable phase.

Methyl Orange

Methyl orange is an acid-base indicater. It is used when a strong acid is titrated against a weak base.

Its indicator range is 3.1 (red) to 4.4 (yellow).

Michaelis-Menten Mechanism

The mechanism of enzyme-catalyzed reaction as suggested by Michaelis and Menten is

$$E+S \underset{k_{-1}}{\overset{k_1}{\rightleftharpoons}} ES$$

$$ES \overset{k_2}{\rightleftharpoons} P+E$$

where E is the enzyme, S the substrate, ES an enzyme complex and P the product. The rate expression of the above mechanism is

$$\left(\frac{d[P]}{dt}\right)_0 = \frac{k_2[E]_0[S]_0}{[S]_0+K_M}$$

where $K_M = (k_{-1}+k_2)/k_1$.

For high $[S]_0$, the rate expression is

$$\left(\frac{d[P]}{dt}\right)_0 = k_2[E]_0$$

that is, the reaction is zero order with respect to S and first order with respect to E.

Miller Indices

Miller indices are used to designate the orientations of different sets of planes in the crystal lattice. These are obtained as follows.

1. Take the reciprocal of Weiss indices. The latter are the ratios of three intercepts h', k' and l' of a plane along the three basis directions and the corresponding unit lengths a, b and c.
2. Take the reciprocal of Weiss indices.
3. Multipy the reciprocals (if necessary) by the smallest number to make them all integers.

The resultant integers are known as Miller indices.

Mitscherlich's law of Isomorphism

Mitscherlich based on his work of phosphates and arsenates postulated the following general law.An equal number of atoms combined in the same manner produce the same crystalline form and that the crystalline form of a compound is independent of the chemical nature of the combined atoms and is determined only by their number and method of combination.

In a simple form, this law can be stated as follows.

Substances which are similar in crystalline form and in chemical properties can usually be represented by similar formulae.

Molality

The term molality (symbol : *m*) is used to express the concentration of a solute in the solution. It is equal to the amount of solute present per kg of the solvent. Its units are mol kg^{-1}.

Molar Conductivity

The molar conductivity (symbol : Λ_m) of an electrolyte may be defined as the conductance of a volume of solution containing one molar mass of a dissolved substance when placed between two parallel electrodes which are at a unit distance apart, and large enough to contain between them the whole solution. Mathematically, it is defined as

$$\Lambda_m = \frac{\kappa}{c}$$

where κ is the conductivity and c is the molar concentration. The units of Λ_m are Ω^{-1} cm^2 mol^{-1}.

Molarity

The term molarity (symbol : *M*) is used to express the concentration of a solute in the solution. It is equal to the amount of solute present per dm^3 of the solution. Its units are mol dm^{-3} which is conveniently represented by the symbol M.

Molar Magnetic Susceptibility

Molar magnetic susceptibility of a medium is given by the expression

$$\chi_m = N_A\left(\alpha_m + \frac{\mu^2{}_m}{3kT}\right)$$

where N_A is Avogadro constant,

α_m is the induced diamagnetic susceptibility per atom/molecule,

μ_m is the magnetic moment carried by each atom/molecule,

k is Boltzmann constant,

and T is kelvin temperature.

The molar magnetic susceptibility can be determined experimentally. knowing χ_m, the value of μ_m can be calculated.

From the value of μ_m, the number of unpaired electrons (n) can be computed using the expression

$$\mu_m = 2\mu_B\sqrt{S(S+1)}$$

where $S = n(\frac{1}{2})$.

Molar Mass

The average mass per unit amount of substance (applicable to both atoms and molecules) of a specified isotopic composition is known as molar mass.

The unit of molar mass is g mol^{-1} or kg mol^{-1}.

Molar Polarization

Molar polarization of a medium is defined as

$$P_m = \frac{\varepsilon_r - 1}{\varepsilon_r + 2}\frac{M}{\rho}$$

where ε_r is the realative permittivity of the medium and M/ρ is the molar volume. The quantity P_m is an additive quantity, and can be calculated if the values of atomic polarization of the constituent atoms in a molecule are known.

Molar Refraction

Molar refraction of a medium is defined as

$$R_m = \frac{n^2 - 1}{n^2 + 2}\frac{M}{\rho}$$

where n is the refractive index and M/ρ is the molar volume of the medium. The quantity R_m is an additive quantity.

Mole

The mole is the amount of substance of a system which contains as many elementary entities as there are atoms in 0.012 kilogram of carbon—12. When the mole is used, the elementary entities must be specified and may be atoms, molecules, ions, electrons, other particles, or specified groups of such particles.

Molecular Mass

The average mass per molecule of a specified isotopic composition of a substance is known as molecular mass. It is simply a mass and thus has the unit of mass (i.e. kg or g). It is given as

mass of a molecule, $M = M_r\ m_{au}$

Molecularity of Reaction

See, elementary step

Molecular Oribital Method

It is one of the methods to explain bonding in the molecule. Its basic postulates are as follows.

1. Each electron in a molecule is described by a wave function, known as the molecular orbital. These molecular orbitals entend over the entire region of the molecule.

2. The square of molecular orbital represents the relative probability of finding an electron in the molecule.

3. Each molecular orbital is represented by drawing contours of constant ψ or ψ^2 or by drawing a boundary surface which includes the majority of charge cloud.

4. Each molecular orbital is associated with a definite energy which may be calculated by solving the appropriate Schrodinger equation.

5. The allocation of electrons to the various molecular orbitals of a molecule is done following the aufbau and Pauli exclusion principles.

One of the methods to construct molecular orbitals is linear combination of valence atomic orbitals centred on the different constituent atoms of the molecule.

Mole Fraction

The mole fraction of a substaece in a mixture of substances is defined as the ratio of amount of the substance to the total amount of substance present in the mixture, i.e.

$$x_i = \frac{n_i}{\sum_i n_i}$$

It is unitless quantity.

Monotropy

If the change of a substance from one polymorphic form to another cannot be reversed, the substance is said to exhibit monotropy. The common example is the transformation of white P to violet P.

Morse Potential

One of the expressions which fits the potential energy curve of a real molecule to a good approximation is due to P.M. Morse and is known as Morse potential. The energy expression is

$$V = D\,[1 - \exp(a(r_{eq} - r))]^2$$

where D is the dissociation energy and a is a constant.

Most Probable Speed

The speed possessed by the maximum fraction of the molecules is known as most probable speed. It is given by the expression

$$u_{mp} = \sqrt{\frac{2RT}{M}}$$

Mutual Solubility Temperature

For any composition of two partially miscible liquids, a minimum definite temperature exists at which the two liquids form a completely miscible liquid. This temperature is known as mutual solubility temperature.

N

Natural Independent Variables

The natural independent variables of different thermodynamic functions are as follows.

$U=f(S, V)$

$H=f(S, p)$

$A=f(T, V)$

$G=f(T, p)$

These follows from the following thermodynamic relations.

$dU=TdS-pdV$

$dH=TdS+Vdp$

$dA=-SdT-pdV$

$dG=-SdT+Vdp$

Nature of Reaction Based on its $\triangle G$ Value

The nature of a reaction whether spontaneous or nonspontaneous can be predicted from the $\triangle G$ value of the reaction. Since $\triangle G=\triangle H-T\triangle S$, four cases given below may be distinguished.

$\triangle H$	$\triangle S$	$\triangle G$	*Comment*
$-$	$+$	$-$	always spontaneous
$+$	$-$	$+$	never spontaneous
$-$ $+$	$-$ $+$	? ?	$\triangle G$ depends upon conditions

The favourable conditions for any process to be spontaneous are (1) decrease in the enthalpy content and (2) increase in entropy of the system.

Nernst Distribution Law

See, distribution law

The distribution law is valid provided (1) the solutions are ideal and dilute, (2) the presence of solute does not change the mutual solubility of the two layers, and (3) the solute has the same molecular mass in both the layers.

Benzoic acid dimerizes in benzene and is practically undissociated in water. The expression of the distributian law in this case is

$$\frac{c(\text{water})}{\sqrt{c(\text{benzene})}} = \text{constant}$$

where c represents the concentration of benzoic acid in the respective solvent.

Nernst Equation

The half-cell potential depends on the concentration and pressure of species appearing in the half-cell chemical equation. It is given by the expression

$$E_{cell} = E^{\circ}_{cell} - \frac{RT}{nF} \ln\left\{\prod_i c_i^{\nu_i}\right\}$$

where n is the number of electrons involved in the cell reaction ν_i is the stoichiometric number of the species and E°_{cell} is the standard potential when the concentration and pressure of the species are 1 mol L^{-1} and 1 atm, respectively.

Neutral Solution

If the concentration of H^+ in a solution is equal to the concentration of OH^-, then the solution is said to be a neutral solution. In terms of K_w, the neutral solution satisfies the expression

$$[H^+] = \sqrt{K_w} \quad \text{or} \quad [OH^-] = \sqrt{K_w}$$

At 25 °C. the above expression becomes

$$[H^+] = 10^{-7}\ \text{mol dm}^{-3} \quad \text{or} \quad [OH^-] = 10^{-7}\ \text{mol dm}^{-3}$$

In terms of pH or pOH, we have

$$\text{pH} = 7 \quad \text{or} \quad \text{pOH} = 7 \text{ at } 25\ ^{\circ}\text{C}$$

Nodal Point

The point at which wave function has zero value is known as nodal point. For example, nonbonding orbital of a diatomic molecule has a nodal point on the bond axis.

Nonbonding Molecular Orbital

If the energy of resultant molecular obital is same as those of atomic energies, the orbital is said to be nonbonding molecular orbital. The addition of electron in this orbital does not stabilize the energy of the molecule.

Nonideal Solutions

If the constituents of a liquid solution do not follow Raoult's law, it constitutes what is known as nonideal solution. Two types of deviations are observed. These are as follows.

1. *Positive deviation*: If the vapour pressures of constituents are greater than those expected from Raoult's law, the solution shows positive deviation from Raoult's law. In this case, the forces of attraction between unlike molecules are weaker than those existing between like molecules.

2. *Negative deviation*: If the vapour pressures of constituents are lower than those expected from Raoult's law, the solution shows negative deviation from Raoult's law. In this case, the forces of attraction between unlike molecules are greater than those existing between like molecules.

Nonprimitive Unit Cell

If in a unit cell, lattice point besides being at the corners is/are also present at the body centre or face centres, then the resultant unit cell is known as nonprimitive unit cell. There are a total of fourteen primitive and nonprimitive unit cells, known as Bravais lattices.

Normalized Wave Function

When the integral

$$\int \psi^* \psi d\tau = 1$$

the wave function is said to be the normalized wave function. The meaning of this integral is that the total probability of finding the electron over the whole of configuration space is equal to one. If the value of the above integrial is not equal to one, the function is said to be non-normalized wave function which can be converted into a normalized wave function by dividing it by the normalization constant $\sqrt{N}$, where

$$N=\int\psi^{*}\psi d\tau$$

Nuclear Magnetic Resonance Spectrum

The potential energy of a magnetic active nucleus (say, proton) depends upon the value of z-component spin quantum number and is given by the expression

$$V=-(\mu_{N}\, gB)\, m_{I}$$

A transition of the nucleus from the lower value of m_I to the next higher value of m_I requires energy equivalent to

$$\triangle V=\mu_{N}\, gB$$

where $\triangle m_I = 1$. This much of energy lies in the radiofrequency rigion (for $B \simeq 1$ T) and hence the abosption of the appropriate radiation would cause transition. This absorption appears as the absorption signal in the nuclear magnetic resonance spectrum.

Numbers of Components

It is the smallest number of independent chemical constituents by means of which the composition of each and every phase can be expressed. The independent chemical constituent is the one whose concentration can be varied independent of other constituents of the system.

O

Ohm's Law

This law states that

$$R=\rho\left(\frac{l}{a}\right)$$

where R is the resistance offered by a conductor of length l and area of cross section a and ρ is known as resistivity (formely as specific resistance).

Onsager Eqnation

Onsager equation relates the molar conductivity at a particular concentration with that at infinite dilution. The expression is

$$\Lambda_m=\Lambda_m^{\infty}-\left(\frac{2.8\times10^6\ z_+z_-q}{(\varepsilon T)^{3/2}\ (1+\sqrt{q})}\Lambda_m^{\infty}+\frac{41.25\ (z_++z_-)}{\eta(\varepsilon T)^{1/2}}\right)\left(\frac{z_++z_-}{2}\right)^{\frac{1}{2}}c^{1/2}$$

where z_+ and z_- are the charge numbers on cation and anion, respectively, ε is the dielectric constant of the soluent, η is the viscosity of the soluent and q is given as

$$q=\frac{z_+z_-\ (\lambda_+^{\infty}+\lambda_-^{\infty})}{(z_++z_-)\ (z_+\lambda_-^{\infty}+z_-\lambda_+^{\infty})}$$

where λ_+^{∞} and λ_-^{∞} are the limiting ionic conductivities of cation and anion, respectively. In a simple form, Onsager equation can be written as

$$\Lambda_m=\Lambda_m^{\infty}-(A\Lambda_m^{\infty}+B)\ c^{1/2}$$

Orbital

The behaviour of an electron in an atom or molecule is represented by a wave function, the form of which can be obtained by solving the appropriate Schrodinger equation. Each wave function which is soluttion of Schrodinger equation is conventionally known as orbital. In an atom, each orbital is characterized by three quantum numbers.

Order of a Reaction

If the rate of a reaction

$$\nu_1A_1+\nu_2A_2 \longrightarrow \nu_3A_3+\nu_4A_4$$

can be written as

$$\text{rate}=k\,[A_1]^a[A_2]^b$$

then a is known as order of the reaction with respect to A_1, b is the order of the reaction with respect to A_2 and $a+b$ is known as the overall order of the reaction. The order of reaction with respect to any species is determined experimentally.

Orthobaric Densities

The densities of the liquid and of saturated vapour in equilibrium with it, are known as orthobaric densities.

Osmosis

When a pure solvent is separated from a solution by a semipermeable membrane, there occurs flow of solvent from pure solvent side to the solution side. This phenomenon is known as osmosis.

Osmotic Pressure

The hydrostatic pressure developed on the solution side which prevents the phenomenon of osmosis further is known as osmotic pressure.

The phenomenon of osmosis can also be prevented by applying either extra pressure on the solution side or reducing pressure on the solvent side. The osmotic pressure of the solution is equal to the pressure which need to be applied on the solution side or need to be reduced on the solvent side. The osmotic pressure of the solution is given by the expression

$$\Pi=cRT$$

where Π is the osmotic pressure and c is the concentration of all species present in the solution.

Ostwald Dilution Law

At infinite dilution, the whole of weak electrolyte is ionized in the solution, i.e. the whole of weak electrolyte is present in the form of ions. This follows from the derivation given below.

$$\begin{array}{ccc} AB \rightleftharpoons & A^+ + & B^- \\ c(1-\alpha) & c\alpha & c\alpha \end{array}$$

$$K_{eq} = \frac{[A^+][B^-]}{[AB]}$$

$$= \frac{(c\alpha)(c\alpha)}{c(1-\alpha)} \equiv \frac{c\alpha^2}{1-\alpha}$$

Assuming $\alpha << 1$, we write

$$\alpha = \sqrt{\frac{K_{eq}}{c}}$$

Hence, as $c \rightarrow 0$, α will go on increasing and in the limit, it will attain the maximum value of one.

Ostwald Viscometer

It is an apparatus used in the measurement of viscosity of a liquid. It consists of U-shaped glass tube, each arm of which includes one bulb of different size. The smaller-sized bulb is at higher level than the other one and is attached to a capillary tube within the arm of the glass tube. A known volume of liquid is taken in the bigger-sized bulb and is sucked into the smaller-sized bulb and the liquid is allowed to flow through capillary tube and is collected back in the bigger-sized bulb. The time taken to flow a volume, V, of liquid through the capillary tube is noted down. This is related to the viscosity of liquid through the Poiseuille's equation

$$\eta = \frac{\pi r^4 pt}{8lV}$$

where l is the length of capillary tube, r is its radius and p is the pressure head because of which liquid flows in the capillary tube. To avoid the measurements of l and r, the relative method is used where water is used as the reference liquid (viscosity known). Comparison of the times of two liquids (water and given liquid) gives the expression

$$\frac{\eta_l}{\eta_w} = \frac{p_l t_l}{p_w t_w}$$

Now $p \propto \rho$ (density), hence

$$\frac{\eta_l}{\eta_w} = \frac{\rho_l t_l}{\rho_w t_w}$$

Knowing, η_w and other values, η_l can be determined.

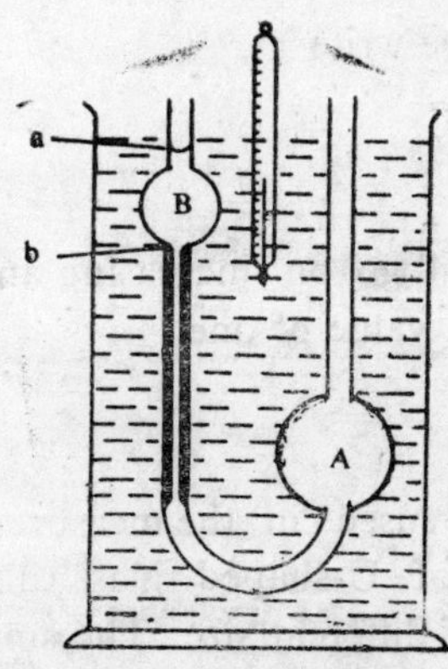

Ostwald viscometer

Overtone Transitions

The transitions of a diatomic molecule from $v=0$ vibrational level to $v=2$, $v=3$, etc. vibrational levels are known as overtone transitions. The transition $v'=2 \leftarrow v''=0$ is known as first overtone, $v'=3 \leftarrow v''=0$ is known as second overtone and so on.

Oxidation-Reduction Half-Cell

An oxidation-reduction half-cell has an inert metal collector, usually platinum, immersed in a solution which contains two ions of the same element in different states of oxidation. One of the examples is ferric-ferrous half cell.

P

Parachor

The physical quantity parachor means comparative volume. It is defined as

$$[P]=V_m\gamma^{1/4}$$

where V_m is the molar volume of the liquid and γ is its surface tension. The comparison of the parachors of different liquids is equivalent to comparison of molar volumes under the conditions of equal (unit) surface tension. The parachor has been found to be largely an additive property and partly a constitutive property.

Paramagnetism

The phenomenon of paramagnetism is due to the presence of unpaired electrons in the atom or molecule, The spinning of electron produces a tiny magnet whose magnetic moment is given by the expression

$$\mu_m=2\mu_B\sqrt{s(s+1)}$$

where μ_B, the Bohr magneton, is given by the expression

$$\mu_B=\frac{eh}{4\pi m_e}$$

and its value is 9.274×10^{-24} J T^{-1}. In the presence of a magnetic field, these tiny dipoles try to align with the magnetic field, and hence lowers the potential energy of interactions.

If an atom or molecule contains more than one unpaired electrons, then

$$\mu_m=2\mu_B\sqrt{S(S+1)}$$

where $S=n(\frac{1}{2})$

where n is the number of unpaired electrons.

Partial Molar Quantities

The partial molar quantity of a substance is defined as

$$Y_{i,pm} = \left(\frac{\partial Y}{\partial n_i} \right)_{t,p,n's \atop j \neq i}$$

where Y is any thermodynamic function. It may be defined in either of the following two ways.

1. It is the change in Y when 1 mole of compound i is added to a system which is so large that the addition has a negligible effect on the composition of the system.
2. Let dY be the change in Y when an infinitesimal amount dn_i of component i is added to a system of definite composition.

The quantity $Y_{i,pm}$ represents the actual value of Y per mole of the ith species in a system of known composition. This value may or may not be equal to the corresponding molar value Y of the species in the pure state. Only for ideal systems, the two values are identical.

Partial Pressure

The partial pressure of a gas in a mixture of gases is defined as the pressure which the gas would exert if it is allowed to occupy the whole volume of the mixture at the same temperature.

Pauli Exclusion Principle

No two electrons in a given atom may have all the four quantum numbers the same.

Within the same orbital, only two electrons can be accommodated. These electrons have opposite spin since only then fourth quantum number (i.e. spin quantum number) has different value.

When the two orbitals are different, the value of at least one of the quantum numbers n, l and m is different and thus the two electrons may have either the same or different spins.

Peptization

Peptization is the process by which certain substances are converted to the colloidal state when shaken with water containing very small amount of an electrolyte, which acts as the peptizing agent. For example, a reddish brown positively charged sol is obtained when freshly precipitated ferric hydroxide is shaken with water containing a small quantity of ferric chloride.

Peritectic Temperature

Peritectic temperature is the melting point at which the liquid obtained has composition different from the the solid phase.

Permittivity of a Medium

The permittivity (symbol : ε) of a medium is defined in terms of Coulomb's law of interaction between charges is

$$F = \frac{1}{(4\pi\varepsilon)} \frac{Q_1 Q_2}{r^2}$$

For a vacuum, the value of permittivity ε_0 is

$$\varepsilon_0 = 8.854 \times 10^{-12} \ \mathrm{C^2 N^{-1} m^{-2}}$$

Phase

A phase is defined as any homogeneous and physically distinct part of a system which is separated from other parts of the system by definite bounding surfaces.

Phase Rule

Phase rule is a general rule which is applicable to a system. This rule states that

$$F + P = C + 2$$

where

F is the degree of freedom of the system,

P is number of phases at equilibrium,

and C is the number of components of the system.

For a reactive system where its constituents can combine with each other, the phase rule is

$$F+P=(C+r)+2$$

where r is the number of independent reactions that are taking place in the system.

pH of Salt Solution

The pH of salt solutions are described below.

(i) A salt formed from a strong acid and a strong base

The cation and anion produced are weak conjugate acid and base, respectively. These do not undergo hydrolysis and hence pH of the solution is the same as that of water, i.e. the resultant solution is neutral.

(ii) A salt formed from a weak acid and a strong base

The anion of such a salt being a strong conjugate base undergoes hydrolysis. Hence, the resultant solution is alkaline in nature. pH of the solution can be computed from the expression

$$\text{pH}=\tfrac{1}{2}\text{p}K_w+\tfrac{1}{2}\,\text{p}K_a+\tfrac{1}{2}\log c$$

where c is the concentration of salt.

(iii) A salt formed from a strong acid and a weak base

The cation of such a salt being a strong conjugate acid undergoes hydrolysis. Hence, the resultant solution is acidic in nature. pH of the solution is given by the expression

$$\text{pH}=\tfrac{1}{2}\text{p}K_w-\tfrac{1}{2}\text{p}K_b-\tfrac{1}{2}\log c$$

(iv) A salt formed from a weak acid and a weak base

Both the cation and anion being strong undergo hydrolysis. The resultant solution may be acidic or alkaline or neutral depending upon the ionization constants of the corresponding weak acid and base. The expression of pH is

$$\text{pH}=\tfrac{1}{2}\,(\text{p}K_w+\text{p}K_a-\text{p}K_b)$$

Phenolphthalein

Phenolphthalein is an acid-base indicator, It is used when a strong or weak acid is titrated against a strong base. Its indicator range is 8.3 (colourless) to 10.0 (red).

Phosphorescent Spectrum

A delayed emission of electronic spectrum is termed as phosphorescent spectrum. This is observed when in the excited state, molecule is switched over from one electronic state to another state which is normally a forbidden state, i.e. the molecule undergoes intersystem crossing from singlet to triplet state or vice versa. In the forbidden state, the molecule comes to the lowest vibration state via radiationless decay. Now since the transition from the forbidden electronic state to the ground state is forbidden, the molecule takes time to undergo this transition. Hence, the recording of spectrum is also delayed.

Photochemistry

The science of photochemistry deals with the study of the effect of radiant energy lying in visible and near ultraviolet region on chemical reactions and with rates and mechanisms by which photochemical reactions proceed.

Photoelectric Effect

When a clean metal plate in a vacuum is irradiated with the light of appropriate frequency, the electrons are emitted. This effect is known as photoelectric effect. It is found that

1. The electrons are emitted instantaneously from a given metal plate when it is irradiated with radiation of frequency equal to or greater than same minimum frequency known as threshold frequency.
2. The kinetic energy of the emitted electrons depends on the frequency of the incident radiation, which is found to increase linearly with the increase of the frequency of the incident radiation.

This effect has been explained by Einstein based on the corpuscular nature of radiation.

Photosensitized Reactions

In many photochemical reactions, the reacting substance does not absorb radiation directly, but acquires the energy from some other light absorbing foreign substance. The latter is known as sensitizer. Mercury and cadmium vapours are often used as the sensitizer. One of the examples of photosensitized reactions is the combination of CO and H_2. The uranyl actinometer is also an example of photosensitized reaction where UO_2^{2+} is used for the decomposition of oxalic acid.

Photostationary State

Absorption of radiation by reactants of a reaction at equilibrium increases the rate of forward reaction without directly affecting the rate of the reverse reaction. The latter is, however, increased due to the increase in the concentration of products. Very soon, a new state called the photostationary state (or photochemical equilibrium) is established where the increase in the rate of forward reaction (due to the absorption of light) becomes equal to the increase in the rate of reverse reaction (due to the enhanced concentration of the products).

pH Scale

For the sake of convenience, the hydrogen ion concentration in a solution is defined in term of pH, defined as

$$pH = -\log\{[H^+]/\text{mol dm}^{-3}\}$$

This definition is purely based on the mathematical operation.

Physical Absorption

If the forces of attraction between adsorbent and adsorbate are of van der Waals type, the adsorption is said to be physical adsorption. This adsorption has following characteristics:

1. predominates at low temperatures.

2. involves low heat of adsorption ($\approx$20 kJ mol^{-1})
3. reversible in nature.
4. involves low activation energy.
5. multilayer adsorption.
6. attains equilibrium very rapidly on changing the temperature and pressure of the system.

Planck's Constant

The Planck's constant is a fundamental constant which has a value of 6.626×10^{-34} J s. This is a proportionality constant between energy of a photon and its frequency, i.e.

$$E \propto \nu$$
$$E = h\nu$$

Polymorphism

Sometimes a substance can exist in more than one crystalline form and each form has its own characteristic vapour pressure curve. Such type of substances are said to exhibit the phenomenon of polymorphism Two types of polymorphism are observed depending upon whether the change from one polymorphic form to the other can be carried out in either directions or only in one direction. The former is known as enantiotropy (Greek: opposite change) and the latter as monotropy (Greek; one change).

p orbital

The wave function having azimuthal quantum number equal to one is known as p orbital.

Point Groups

When the seven crystal systems are studied from the view point of the three symmetry combinations (namely, proper rotation axis, mirror plane, and rotation-inversion axis), it is found that the crystals can be classified into 32 different groups. These are known as 32 point groups. The details of these amongst seven crystal systems are as follows,

Triclinic	2
Monoclinic	3
Orthorhombic	3
Trigonal	5
Cubic	5
Tetrogonal	7
Hexagonal	7
Total	32

Poiseuille Equation

The rate of flow through a cylindrical capillary tube of known internal diameter is given by Poiseuille equation:

$$\eta=\frac{\pi r^4 pt}{8lV}$$

where t is time taken for a volume V to pass through a capillary tube of length l and radius r, p is the pressure difference due to which the fluid moves and η is the coefficient of viscosity of the fluid.

Potential Energy of a Nucleus in a Magnetic Field

A nucleus when placed in a magnetic field has potential energy which is given by the expression

$$V=-(\mu g_N m_I)B$$

where m_I is the z-component of spin quantum number of the nucleus which has values of I, $I-1$, .. , $-I+1$, and $-I$, g is the nuclear splitting factor, μ_N is the nuclear magneton and B is the magnetic flux density.

For a proton, the permitted values of m_I are $+\frac{1}{2}$ and $-\frac{1}{2}$. Hence, protons will have two types of potential energy depending upon the value of m_I equal to $+\frac{1}{2}$ and $-\frac{1}{2}$, respectively.

Potentiometric Titrations

The potential measurements provide one of the most important analytical techniques for carrying out acid-base, redox and precipitation titrations. The equivalence point of titrations can be determined accurately without making use of any visual indicators. The principle underlying the potentiometric titrations is the Nernst equation according to which the potential of an electrode depends upon the concentrations of reduced and oxidized forms of the involved redox couple. During the titration, the concentrations of these species vary, which in turn, varies the potential of the electrode. The variation of potential is gradual in the beginning but becomes faster as the equivalence point is reached. The variation is maximum at the equivalence point and becomes lesser and lesser beyond the equivalence point. Thus, at the equivalence point, one observes a point of inflexion.

The equivalence point of titration is determined graphically. A typical plot of cell potential versus volume of titrant added is shown in the following figure.

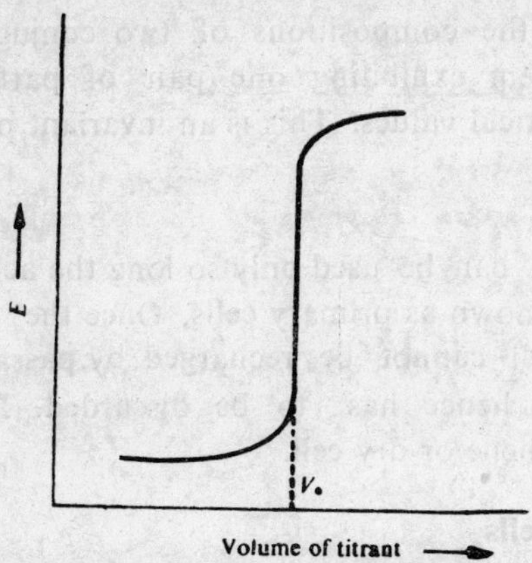

The equivalence point of titration which corresponds to the point of inflexion can be determined from the plot of E vesus V. A more exact method is to plot $\triangle E/\triangle V$ versus V. The maximum on the curue corresponds to the equivalence point of the titration. One may also plot $\triangle^2 E/\triangle V^2$ to determine the equivalence point. These are shown in the following figures.

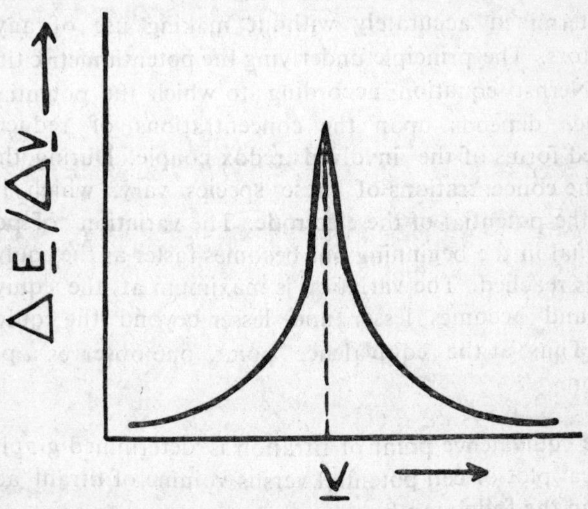

Plait Point

At plait point, the compositions of two conjugate solutions of a ternary system exhibiting one pair of partially miscible liquids have identical values. This is an invariant point.

Primary Cell

The cells which can be used only so long the active materials are present are known as primary cells. Once the materials are consumed, the cell cannot be recharged by passage of current through it, and hence has to be discarded. The common example is Leclanche or dry cell.

Primitive Unit Cells

If in a unit cell, lattice points are present only at the corners, then the resultant unit cell is known as primitive unit cell. Based on the symmetry of unit cells, these can be classified into seven categories, known as seven crystal systems. *See* also crysal systems, p. 34.

Principle of Le Chatelier and Braun

This principle predicts qualitatively the effects on the system at equilibrium when some of the variables such as temperature, pressure and concentration are changed. It states as follows.

If a system at equilibrium is subjected to a change, the system will react in such a way so as to oppose or reduce the change if this is possible, i.e. the system tends to balance or counteract the effects of any imposed stress.

Examples of this principle are:

1. Pressure increase	Reaction will shift in a direction where the number of gaseous molecules is reduced thus lowering the pressure *p*.
2. Heat added or temperature is increased	The equilibrium will shift in the endothermic direction, i.e. it is shifted to high enthalpy side.
3. One of the components of the system is added	Reaction proceeds in a direction so as to reduce the amount of this component.

Principal Quantum Number

This number (symbol: *n*) describes the energy of the electron in the hydrogen like systems and is given by the expression

$$E = -\frac{1}{n^2}\left[\frac{2\pi^2 m_e Z^2 (e\sqrt{4\pi\varepsilon_0})^4}{h^2}\right]$$

where the various symbols have their usual meaning.

The permitted values of *n* are 1, 2, 3,....

Quantization of Rotational Energy

The rotational energy of a molecule are quantized. For a

diatomic molecule, the expression is

$$E=J\ (J+1)\ \frac{h^2}{8\pi^2\ I}$$

where J is the rotational quantum numbeer $(=0,\ ',\ 2,\ldots)$ and I is the moment of inertia of molecule $(=\mu r^2)$.

Quantization of Translational Energy

The translational energies of a particle moving in one direction are quantized and are given by the expression

$$E = n^2 \left[\frac{h^2}{8l^2\ m} \right]$$

where n is quantum number which can have integral values 1,2,3,..., l is the length of the box in which the particle of mass m moves and h is Planck's constant.

In a three dimensional box, the energy expression is

$$E = \frac{h^2}{8m} \left[\frac{n_1^2}{l_1^2} + \frac{n_2^2}{l_2^2} + \frac{n_3^2}{l_3^2} \right]$$

where n_1, n_2 and n_3 are the quantum numbers, each having integral values and l_1, l_2, and l_3 are the three lengths of the sides of box.

The value of $\triangle E$ between the two translational levels is too small to be determined experimentally and for all practical purposes, the translational energy may be considered to be continuous.

Quantization of Vibrational Energy

The vibrational energies of molecules are quantized. For a diatomic molecule, the expression is

$$E=(v+\tfrac{1}{2})\ h\ v_o$$

where $v_o = \frac{1}{2\pi}\sqrt{\frac{k_f}{\mu}}$

where k_f is the force constant of the molecule. It is a measure of rigidity of the bond existing in the molecule, and μ is the reduced mass of the molecule which is given by the expression

$$\frac{1}{\mu}=\frac{1}{m_1}+\frac{1}{m_2}$$

where m_1 and m_2 are the masses of two atoms of the diatomic molecule. The symbal v stands for vibrational quantum number and it can have values of 0,1,2....

Quantum Efficiency

The quantum efficiency of a photochemical reaction is defined as

$$\phi=\frac{\text{Number of molecules reacted}}{\text{Number of photons absorbed}}$$

In terms of amount of substance, we have

$$\phi=\frac{\text{Amount of substance reacted}}{\text{Amount of photons absorbed}}$$

For a primary process (i.e. light absorbing process), the quantum efficiency has a value of one. The symbol ϕ stands for the overall quantum efficiency of the reaction.

Quantum Numbers

In solving Schrödinger equation for hydrogen atoms, three quantum numbers are automatically emerged. The physical significance of these numbers are as follows.

Principal quantum number: See p. 125.
Azimuthal quantum number: See p. 12
Magnetic quantum number: See p. 99.

Quinhydrone Electrode

Quinhydrone is the name given to the molecular crystal formed between quinone (Q) and hydroquinone (QH_2). When dissolved in water, the crystal decomposes into its chemical compounds. The reduction reaction is

$$Q+2H^++2e^-\rightarrow QH_2$$

Its cell potential is

$$E=E^\circ-\frac{RT}{2F}\ln\frac{1}{[H^+]^2}$$

with $E^\circ=0.6996$ V.

The quinhydrone electrode is commonly used to find the pH of unknown solution.

R

Radioactive Decay

The spontaneous emission of α-particles ($_2{}^4He^{2+}$), β-particles (electrons) and γ-radiations from a heavy atom is known as radioactive decay.

Raman Spectroscopy

Raman spectroscopy deals with the indirect changes in the vibrational and rotational levels of a molecule. The sample is exposed to visible radiation (which does not correspond to the electronic changes in the molecule) and the scattered radiation is analyzed. The analysis indicates that the scattered radiations not only include incident radiation but also other radiations with wavelengths greater and smaller than the incident radiation. These radiations are due to the addition or subtraction of energies corresponding to the vibrational changes or/and rotational changes in the molecule to the energy of incident radiation. These have very weak intensities. Only those vibrations and rotations are Raman active for which molecular polarizability changes during the vibrations and rotations. Raman spectrum provides the useful informations which sometimes are not provided by direct spectroscopy.

Ramsay and Shields Equation

At temperature not too near the critical point, the molar surface energy variation with temperature is given as

$$\gamma(Mv)^{2/3}=k\ (t_c-t-6)$$

Raoult's Law

The vapour pressure of the solution containing a nonvolatile solute is equal to the vapour pressure of the pure solvent multiplied by the mole fraction of the solvent in the solution. Mathematically, it is written as

$$p_1 = x_1 p_1^*$$

where the subscript 1 stands for solvent.

Rate Constant

If the rate of a reaction

$$\nu_1\ A_1 + \nu_2\ A_2 \longrightarrow \nu_3\ A_3 + \nu_4\ A_4$$

can be written as

$$\text{rate} = k\ [A_1]^a\ [A_2]^b$$

where k is a constant known as rate constant. It may be defined as the rate of reaction when the concentrations of species involved in the rate equation are unity. Its units are (mol $dm^{-3})^{1-\gamma} s^{-1}$, where γ is the overall order of the reaction. The rate constant of a reaction is always determined experimentally.

Rate of Reaction

In chemical kinetics, rate of a reaction is uniquely defined as the rate of change of degree of advancement of the reaction. It has the units of mol s^{-1}. Taking a general reaction, we can write

	ν_1A_1	+	ν_2A_2	$\longrightarrow$	ν_3A_3	+	ν_4A_4
$t=0$	n_1		n_2		0		0
$t=t$	$n_1-\nu_1\xi$		$n_2-\nu_2\xi$		$\nu_3\xi$		$\nu_4\xi$

where ξ is the degree of advancement of the reaction. Now

$$-\frac{dn(A_1)}{dt} = -\frac{d\,(n_1-\nu_1\xi)}{dt} = \nu_1\frac{d\xi}{dt}$$

$$-\frac{dn(A_2)}{dt} = -\frac{d(n_2-\nu_2\xi)}{dt} = \nu_2\frac{d\xi}{dt}$$

$$\frac{dn(A_3)}{dt} = \frac{d(\nu_3\xi)}{dt} = \nu_3 \frac{d\xi}{dt}$$

$$\frac{dn(A_4)}{dt} = \frac{d(\nu_4\xi)}{dt} = \nu_4 \frac{d\xi}{dt}$$

Obviously

$$r = \frac{d\xi}{dt} = -\frac{1}{\nu_1}\frac{dn(A_1)}{dt} = -\frac{1}{\nu_2}\frac{dn(A_2)}{dt}$$

$$= \frac{1}{\nu_3}\frac{dn(A_3)}{dt} = \frac{1}{\nu_4}\frac{dn(A_4)}{dt}$$

Thus, the rate of reaction is defined as the rate of change of the amount of a reactant or a product divided by the corresponding stoichiometric number appeared in the balanced chemical equation.

Rate of Reaction Divided by Volume

In chemical kinetics, it is more useful to use the term rate of reaction divided by specified volume. This volume may be dependent or independent of time and may or may not be that of a single phase in which the reaction is taking place. For example, for the reaction

$$\nu_1 A_1 + \nu_2 A_2 \longrightarrow \nu_3 A_3 + \nu_4 A_4$$

we have

$$r = \frac{d\xi/dt}{V} = \frac{1}{V}\frac{1}{\nu_i}\frac{dn_i}{dt}$$

where subscript i refers to any one reactant or product. If volume is independent of time, we can write this expression as

$$r = \frac{1}{\nu_i}\frac{dn_i/V}{dt} = \frac{1}{\nu_i}\frac{dc_i}{dt}$$

where c_i is the concentration of ith component (reactant or product) of the reaction. This units of rate of reaction divided by volume (also conventionally known as rate of reaction) is mol dm^{-3} s^{-1}.

Reaction Potential

$$\triangle\tilde{G}=\left[\frac{\partial G}{\partial \xi}\right]_{T,\,p}$$

The value of $\triangle\tilde{G}$ is equal to the free energy change of the reaction and is given by the expression

$$\triangle\tilde{G}=\sum_{j}\nu_{j}\mu_{j}$$

where ν_js are the stoichiometric numbers and μ_js are the chemical potentials.

Reaction Potential and Reaction Quotient

The reaction potential ($\triangle\tilde{G}$) and reaction quotient (Q_p) are related to each other through an expression

$$\triangle\tilde{G}=\triangle\tilde{G}^{\circ}+RT\ln Q_p$$

where $\triangle\tilde{G}^{\circ}$ is the standard reaction potential of the reaction.

Reaction Quotient

The reaction quotient of a gaseous reaction is given by the expression

$$Q_p=\frac{\text{Product of pressures of products each raised to the corresponding stoichiometric number}}{\text{Product of pressures of reactants each raised to the corresponding stoichiometric number}}$$

Mathematically, it is written as

$$Q_p=\Pi_i\,(p_i)^{\nu_i}$$

where ν_i are the stoichiometric numbers which have positive values for the products and negative values for the reactants, and the symbol Π stands for multiplication sign.

If each pressure is divided by the standard unit pressure, one gets the standard reaction quotient (symbol: $Q_p°$)

It is defined as

$$Q_p° = \prod_i (p_i/p°)^{\nu_i}$$

where $p°$ is 1 atm. Taking an example of

$$2HCl(g) + \tfrac{1}{2}\, O_2(g) \longrightarrow H_2O(g) + Cl_2(g)$$

$$Q_p° = \frac{(p_{H_2O}/p°)\;(p_{Cl_2}/\;p°)}{(p_{HCl}/p°)^2\;(p_{O_2}/\;p°)^{1/2}}$$

Reduced Equation of State

See, law of corresponding states.

Reference Half-Cells

The hydrogen half-cell with respect to which the potential of other half-cell is determined is very inconvenient to use. To avoid this, reference half-cells whose potentials have been determined accurately are used. The most common type of reference half-cell has the following form.

saturated solution of sparingly soluble salt of metal	+	additional strongly ionized salt with a common ion	|	Metal

Examples include

1. Calomel electrode $Cl^-(aq) \mid Hg_2Cl_2(s) \mid Hg$
2. Silver-silver chloride electrode $Cl^-(aq) \mid AgCl(s) \mid Ag$
3. Merrury-mercurous sulphate electrode $SO_4^{2-}(aq) \mid Hg_2SO_4(s) \mid Hg$

Relation Between $\triangle H$ and $\triangle U$

The expression connecting $\triangle H$ and $\triangle U$ of a chemical equation is

$$\triangle H = \triangle U + (\triangle \nu_g)RT$$

where $\triangle\nu_g$ is the change in the stoichiometric numbers of gaseous species in going from reactants to products and is given by the expression

$$\triangle\nu_g = \Sigma\nu_g(\text{product}) - \Sigma\nu_g(\text{reactant})$$

Relation Between K_p and K_c

K_p is expressed in terms of pressure units and K_c is expressed in terms of concentration units. The two are related to each other through an expression

$$K_p = K_c\ (RT)^{\triangle\nu_g}$$

The expression connecting K_p° and K_c° is

$$K_p^\circ = K_c^\circ \left[\frac{c^\circ\ RT}{p^\circ}\right]^{\triangle\nu_g}$$

where $c^\circ = 1$ mol dm^{-3} and $p^\circ = 1$ atm.

Relation Between K_p and K_x

The expression connecting K_p and K_x is

$$K_p = K_x\ (p_{total})^{\triangle\nu_g}$$

Ths expression connecting corresponding standard quantities is

$$K_p^\circ = K_x^\circ \left[\frac{p_{total}}{p^\circ}\right]^{\triangle\nu_g}$$

where p° is 1 atm.

Relative Atomic Mass of an Element

The ratio of the average mass per atom of a specified isotopic composition of an element to 1/12 of the mass of an atom of nuclide carbon-12, i.e.

$$A_r = \frac{\text{Mass of an atom}}{(1/12)\text{ mass of an atom of }^{12}\text{C}}$$

The relative atomic mass carries no units as it is the ratio of two masses.

Relative Lowering of Vapour Pressure

By definition, the relative lowering of vapour pressure is written as $-\Delta p/p^*$, where $-\Delta p$ is the lowering of vapour pressure of a solvent when a nonvolatile solute is dissolved in it and p^* is the vapour pressure of pure solvent. According to Raoult's law, this is given as

$$-\frac{\Delta p}{p^*}=-\frac{p-p^*}{p^*}$$

$$=-\frac{x_1 p^*-p^*}{p^*}=1-x_1$$

$$=x_2$$

that is, relative lowering of vapour pressure is equal to the mole fraction of solute in the solution.

Relative Molecular Mass of a Substance

The ratio of the average mass per molecule of a specified isotopic composition of a substance to 1/12 of the mass of an atom of the nuclide ^{12}C, i.e.

$$M_r=\frac{\text{Mass of a molecule}}{(1/12)\text{ mass of an atom of }^{1\cdot}\text{C}}$$

The relative molecular mass is a pure number (i.e. it carries no units) as it is the ratio of two masses.

Relative Permittivity

The ratio ε_r defined as

$$\varepsilon_r=\frac{\varepsilon}{\varepsilon_0}$$

is known as relative permittivity. In CGS units, the relative permittivity is known as dielectric constant.

Resistance

The medium, through which current or electricity is flowing, exerts obstruction to the passage of current. This obstruction is known as resistance and is expressed in ohm unit (symbol: Ω).

The expression connecting the current, (I), potential difference ($\triangle\phi$) and resistance (R) is

$$I=\frac{\triangle\phi}{R}$$

One volt is required to force a current of one ampere through a resistence of one ohm

Resistivity

Resistivity is the resistance offered by a conductor of unit length and unit area of cross section. Its units are ohm cm or ohm m.

Reversible Elementary Reactions

Reversible elementary reactions are those reactions in which both forward and backward reactions take place simultaneously.

Rock Salt Structure

In rock salt structure, anions form cubical-closest packing and cations occupy octahedral holes.

Root Mean Square Speed

It is defined as the square root of the average of the squares of speeds of gaseous molecules in a system. Mathematically, it is written as

$$\sqrt{\overline{u^2}}=\sqrt{\frac{u_1^2+u_2^2+\ldots+u_N^2}{N}}$$

It can be shown that this is given by the expression

$$\sqrt{\overline{u^2}}=\sqrt{\frac{3RT}{M}}$$

Rotational Spectroscopy of Diatomic Molecules

The rotational energies of diatomic molecule is quantized and are given by the expression

$$E=\frac{h^2}{8\pi^2 I}J(J+1)$$

where I is the moment of inertia of the molecule and J is the rotational quantum number (=0, 1, 2, 3,...). The expression of $\triangle\tilde{E}$ between the two neighbouring levels is

$$\triangle\tilde{E}=2B\,(J''+1)$$

where B, rotational constant $(=h/8\pi^2Ic)$, is expressed in wave number unit and J'' has integral value (=0,1,2,...).

For diatomic molecules, $\triangle E$ lies in the radiofrequency region and the absorption of radiation of appropriate frequency causes rotational transitions. The actual rotational spectrum includes many absorptions with spacing equal to $2B$. Many absorptions are due to the fact that many of rotational levels are occupied by molecules because the thermal energy $kT >> \triangle E$. The rotational spectrum provides the value of B of the molecule which is, in turn, used to calculate internuclear distance of the molecule.

Rule of Mutual Exclusion

If a molecule has a centre of symmetry, then Raman active vibrational modes are infrared inactive, and vice versa. This rule is known as rule of mutual exclusion.

Rydberg Constant

The Rydberg constant is given by the expression

$$R=\frac{2\pi^2e^4m}{h^3c}$$

where m is the mass of electron, e is its charge, h is planck's constant and c is the speed of light. The value of R is

1.09737×10^5 cm^{-1}.

Salt Bridge

It is a devise to connect the two half cells of a galvanic cell. It is required to maintain the electrical neutrality in the two half cells. It is a U-type glass tube which is filled with an agar jelly saturated with either KCl or NH_4NO_3.

Schrödinger Wave Equation

The description of behaviour of electron must be compatible with de-Broglie relation and uncertainty principle. In the light of this, Schrödinger proposed an equation for the behaviour of electron in an atom. The equation is

$$\nabla^2\psi + \frac{8\pi^2 m}{h^2}(E-V)\,\psi = 0$$

where ∇^2 is Laplacian operator given as

$$\nabla^2 = \frac{\partial^2}{\partial x^2} + \frac{\partial^2}{\partial y^2} + \frac{\partial^2}{\partial z^2}$$

and V is the expression of classical potential energy of the electron in an atom. Solving the above equation for ψ describes the complete behaviour of electron.

Second

The second is the duration of 9192631770 periods of the radiation corresponding to the transition between the two hyperfine levels of the ground state of the caesium—133 atom.

Secondary Cells

The cells which can be used again and again by recharging the active materials are known as secondary cells. The common example of this type of cell is lead storage cell.

Second Law of Thermodynamics

See Clausius statement of second law of thermodynamics and Kelvin-Planck statement of second law of thermodynamics.

Second-Order Phase Transition

In the second-order phase transition, the chemical potential and its derivatives are continuous functions of temperature whereas the second derivatives are discontinuous. The continuity of first derivative implies that the entropy, enthalpy and volume of the system do not change when the transition occurs.

Examples of this type of transitions are order-disorder transitions in alloys and the fluid-superfluid transition in helium.

Selection of an Acid-Base Indicator

The selection of an acid-base indicator depends on the pH of the solution at its equivalence point.

For *strong acid* (HCl) *versus strong base* (NaOH), the salt (NaCl) formed at the equivalence point does not undergo hydrolysis and hence the pH of the solution is 7. With a slight excess base, pH goes in the region of 8 to 10. Hence, phenolphthalein with pink colour as the end point is used as the indicator.

For *weak acid* (acetic acid) versus *strong base* (NaOH), the salt (sodium acetate) formed at the equivalence point under goes hydrolysis generating free OH^-. Consequently, the pH of the solution at the equivalence point is about 8 and with a slight excess of base, it is shifted to 9 to 10. Hence, phenolphthalein with pink colour as the end point is used as the indicator.

For *weak base* (ammonium hydroxide) versus *strong acid* (hydrochloric acid), the salt (ammonium chloride) undergoes hydrolysis generating free H^+. Consequently, the pH of the solution at the equivalence point is about 6 and with a slight excess of acid, it becomes still less (say, 3 to 4). Hence, methyl orange with red colour as end point is used as the indicator.

Semipermeable Membrane

A membrane which allows only the passage of solvent molecules through it and not those of solute molecules is known as semipermeable membrane.

Solubility Product

For a sparingly soluble salt such as AgCl, the following equilibrium exists:

$$AgCl(s) \rightleftharpoons Ag^+(aq) + Cl^-(aq)$$

Its equilibrium constant is given by the expression

$$K_{eq} = \frac{[Ag^+][Cl^-]}{[AgCl]}$$

Since, [AgCl] remains constant, a new equilibrium constant, known as solubility product, can be defined as

$$K_{sp} = K_{eq}[AgCl] = [Ag^+][Cl^-]$$

Thus, solubility product of a sparingly soluble salt is the product of its ionic concentrations present in the solution.

For a salt containing unequal numbers of cations and anions, each concentration term is raised to a power equal to the corresponding stoichiometric number appeared in the chemical equation. For example, for $Ca_3(PO_4)_2$, the solubility product is defined as

$$K_{sp} = [Ca^{2+}]^3[PO_4^{3-}]^2$$

Shape of an Orbital

The region around the nucleus where there exists 90–95% probability of its existence is conventionally known as shape of orbital.

1. s orbitals are spherical in nature.

2. p orbitals are dumb-bell shaped. These are three in number.

3. f orbitals are double dumb-bell shaped. These are five in number.

SI Convention of Heat and Work

According to SI convention, heat absorbed by the system and

work done on the system are taken as positive quantities as these operations increase the internal energy of the system. On the other hand, heat released from a system and work done by the system are taken as negative quantities as there operations decrease the internal energy of the system. In accordance with the SI convention, the expression of infinitesimal work during the volume change of the system is given by

$$dw = -p_{ext} dV$$

SI Units

In order to have consistency in the scientific recording, IUPAC (international union of pure and applied chemistry) has recommended the use of coherent units, known as the international system of units commonly abbreviated as SI (Systeme International d'Unites).

The SI base units of the seven independent physical quantities are as follows.

Physical quantity	*Name of SI unit*	*Symbol*
length	metre	m
mass	kilogram	kg
time	second	s
electric current	ampere	A
thermodynamic temperature	kelvin	K
amount of substance	mole	mol
luminous intensity	candela	Cd

Significance of Standard Half-Cell Potentials

The tabulated values of half-cell potentials are relative to that of $H^+ \mid H_2 \mid Pt$ electrode and these are the reduction potentials. By convention, $E°(H^+ \mid H_2 \mid Pt) = 0$ at 25 °C. A positive potential of another half-cell means, the tendency of reduction reaction of this half-cell is more than that of H^+ into H_2. A negative value means the tendency of half-cell is less than that of H^+ into H_2.

Singlet and Triplet States

If all electrons in an atom or a molecule are paired, this leads to the singlet state since only one wave function is sufficient to define to system completely. It two electrons in an atom or a molecule are unpaired, this leads to the triplet state as three wave functions are required to describe the three different spin states of the system. The triplet state has lower energy than the corresponding singlet state.

Solid Solution

A solid solution is formed when one solid dissolves into another to give one perfectly homogeneous solid phase. The space lattice of solid solution reveals that one constituent enters the lattice of the other and is uniformly distributed throughout.

Solute

In a solution, the component present in lesser amount is referred to as solute.

Solvent

In a solution, the component present in larger amount is referred to as solvent.

Solvent Extraction

The technique of extraction of a solute in a solvent by another solvent is named as solvent extraction. When a solution containing the solute is shaken with another immiscible solvent, the solute distributes between the two solutions according to the Nernst distribution law. For efficient use of solvent, it is advantageous to carry out this process in a series of successive stages.

The fraction of solute remaining unextracted after the *n*th successive step of addition volume V_1 of the solvent 2 in the volume V_2 of the solution containing the solute is

$$\frac{w_n}{w}=\left(\frac{K_d\ V_1}{K_d\ V_1+V_2}\right)^n$$

If the entire solvent 2 is added in one step, then the fraction of solute remaining unextracted is

$$\frac{w'}{w}=\left(\frac{K_d\ V_1}{K_d\ V_1+nV_2}\right)$$

It can be shown that

$$w_n < w'$$

that is, the amount of solute remaining unextracted in multistage extraction is smaller than that in a single stage extraction.

s Orbital

The wave function having azimuthal quantum number equal to zero is known as s orbital.

Space Groups

If the crystals are studied from the view point of translational symmetry as well as point symmetry, these can be classified into 230 groups, known as space groups. The details of these along the seven crystals systems are as follows.

Triclinic	2
Monoclinic	13
Orthorhombic	59
Tetragonal	68
Hexagonal	27
Trigonal	25
Cubic	36
Total	230

Space Lattice

The regular three-dimensional arrangements of identical points in space gives rise to a space lattice. The definition of a space lattice is strictly a geometrical concept and represents a three-dimensional translational repetition of the centres of gravity of the units of pattern in the crystal.

Spin Qnantum Number

An electron in an orbital can also spin around its own axis. This spin is characterzed by spin quantum number s, such that the angular momentum of the electron due to the spin is given by the expression

$$L = \sqrt{s(s+1)} \left(\frac{h}{2\pi}\right)$$

The permitted value of s is $\frac{1}{2}$.

Two types of spin, clockwise and anticlockwise, are possible. These are characterized by the constant m_s which has a value of $+\frac{1}{2}$ or $-\frac{1}{2}$. The z-component of angular momentum due to spin is given by the expressions

$$\alpha = spin : m_s = \tfrac{1}{2} \quad , \quad L_z = \tfrac{1}{2} \left(\frac{h}{2\pi}\right)$$

$$\beta = spin : m_s = -\tfrac{1}{2} \, , \, L_z = -\tfrac{1}{2} \left(\frac{h}{2\pi}\right)$$

Spin Quantum Number of a Nucleus

The spinning of a nucleus is described by spin quantum number which has the following values depending upon the nucleus.

1. Nuclei having both charge and mass even have zero spin quantum number. Examples include $^{4}_{2}He$, $^{12}_{6}C$ and $^{16}_{8}O$.
2. Nuclei having odd charge and even mass have integral spin quantum number. Examples include, ^{2}H, ^{14}N (spin = 1) and ^{10}B (spin 3).

3. Nuclei with odd mass have half-integral spin. Examples include ^1H, ^{15}N (spin=1/2), ^{17}O (spin=5/2), ^{19}F (spin =1/2) and ^{35}Cl (spin=3/2).

Spin-Spin Interaction

The actual field seen by a magnetic active nucleus (say, proton) gets affected by the number of neighbouring magnetic active nuclei. This fact is known as spin-spin interactions. Its net effect is that a particular nucleus shows more than one absorption in the nuclear magnetic resonance spectrum depending upon the number and nature of neighbouring magnetic active nuclei.

Stalagmometer

The surface tension of a liquid can be determined by using the apparatus known as stalagmometer. It consists of a bulb attached to a fine capillary with a sharp edge. The liquid is allowed to fall drop by drop through the capillary tube. The mass of drop detaching from the capillary tube is related to the surface tension of the liquid through the expression

$$mg=(2\ \pi r)\gamma$$

where r is the radius of capillary tube. To avoid the measurement of r, relative method is used where water is used as the reference liquid.

Standard Cell Potential

The absolute value of the standard half-cell potential cannot be determined. However, its value relative to hydrogen electrode can be determined experimently. By convention, the standard potential of hydrogen electrode (where pressure of hydrogen gas is 1 atm and concentration of H^+ is 1 mol L^{-1}) has been assigned a value of zero at 25 °C. The potential of the cell determined this way is known as standard cell potential.

Standard Enthalpy Changes of a Chemical Equation

The standard enthalpy change of a chemical equation can

be determined by consulting the table of standard molar enthalpies. The expression to be used is

$$\triangle H^\circ = \sum_{\text{products}} \nu_i H_i^\circ - \sum_{\text{reactants}} \nu_i H_i^\circ$$

where ν_i's are the corresponding stoichiometric coefficient apppeared in the balanced chemical equation.

The units of $\triangle H^\circ$ are kJ mol^{-1}.

Standard Equilibrium Constant and Standard Cell Potential

The standard equilibrium constant of a reaction is related to the standard free energy of the reaction by the expression

$$\triangle G^\circ = -RT \ln K^\circ_{eq}$$

Now,

$$\triangle G^\circ = -nFE^\circ$$

Hence,

$$nFE^\circ = RT \ln K_{eq}^\circ$$

or

$$K_{eq}^\circ = \text{antilog}\left(\frac{nFE^\circ}{RT}\right)$$

Standard Free Energy of Formation

The standard free energy of formation of a substance is the free energy change of an equation in which 1 mol of the compound in its standard state is formed from elements in their standard states. By convention, the free energy of formation of the elements is considered to be equal to zero.

In aqueous solution, the standard state is unit concentration. Free energy of formation of ionic species are defined by the convention that the standard free energy of H^+ is zero.

Standard Gibbs Function

The standard Gibbs function (symbol; G°) of a substance at a given temperature is the Gibbs function of the substance at

1 atm pressure. The standard Gibbs function depends only on the temperature of the system. For an ideal gas

$$G_{T,p}=G_T{}^{\circ}+RT \ln \left(\frac{p}{1 \text{ atm}}\right)$$

Standard Molar Entropy

The standard molar entropy of a substance is the molar entropy when the pressure is 1 atm. (symbol: S°). For an ideal gas, the expression of molar entropy at pressure other than 1 atm is given by the expression

$$S_m=S_m{}^{\circ}-R \ln \left(\frac{p}{1 \text{ atm}}\right)$$

Standard Reaction Potential

The change in standard Gibbs function of a reaction is defined as

$$\triangle \tilde{G}^{\circ}=\sum_j \nu_j \mu_j{}^{\circ}$$

where ν_j's are the stoichiometric numbers and $\mu_j{}^{\circ}$ are the standard chemical potential.

Standard Reaction Potential and Standard Equilibrium constant

For the reaction at equilibrium, the reaction potential $\triangle \tilde{G}$ has a zero value. Hence, the expression

$$\triangle \tilde{G}=\triangle \tilde{G}^{\circ}+RT \ln Q_p{}^{\circ}$$

is reduced to

$$\triangle \tilde{G}^{\circ}=-RT \ln K_p{}^{\circ}$$

where at equilibrium

$$Q_p{}^{\circ}=K_p{}^{\circ}.$$

State Function

A function is said to be a state function if its differential satisfies Euler's reciprocity relation. In other words, the function ϕ represents a state function if it is possible to write an expression

$$d\phi = \left(\frac{\partial \phi}{\partial x}\right)_y dx + \left(\frac{\partial \phi}{\partial y}\right)_x dy$$

with the condition that

$$\left\{\frac{\partial}{\partial y}\left(\frac{\partial \phi}{\partial x}\right)_y\right\}_x = \left\{\frac{\partial}{\partial x}\left(\frac{\partial \phi}{\partial y}\right)_x\right\}_y$$

The change in the value of state function depends only on initial and final stages and not on the path of the process carried out in going from initial to final stage. All thermodynamic functions are state functions.

State Variables

The macroscope properties (such as pressure, volume, temperature, mass, etc.) which are stated to describe the state of a system are known as state variables.

Statistical Thermodynamics

This branch of science is based on statistical mechanics and deals with the calculation of thermodynamic properties of matter from the classical or quantum mechanical behaviour of a large congregation of atoms or molecules.

Steam Distillation

The process of steam distillation of immiscible liquids is generally utilized for the purification of those liquids which either boil at too high temperatues or decompose when heated to their normal boiling points. If water is one of the components, then the distillation is known as steam distillation.

The masses of the constituents of a mixture of two immiscible liquid pair in the distillate is given by the expression

$$\frac{m_A}{m_B} = \frac{M_A \, p^*_A}{M_B \, p^*_B}$$

where M_A and M_B are the respective molar masses and p^*_A and p^*_B are the respective vapour pressures of the pure components.

Stern-Gerlach Experiment

The experimental demonstration for the existence of the electron spin is due to Stern-Gerlach. They passed a beam of silver atoms through a strong inhomogeneous magnetic field and found that the beam splits symmetrically in two directions. This is due to α- and β- spinning of the electron in silver atom.

Stern-Volmer Relation

The intensity of fluorescene exhibited by a photochemical system is affected by the use of an appropriate quincher. A kinetic expression for the quenching of fluorescene is obtained by condsidering the scheme given below.

Absorption $\quad M + h\nu \xrightarrow{I_{abs}} M^*$

Fluorescene $\quad M^* \xrightarrow{k_f} M + h\nu'$

Quenching $\quad M^* + Q \xrightarrow{k_q} M + Q + \text{kinetic energy}$

The expression of quenching, which is known as Stern-Volmer relation, is

$$\frac{1}{I_f} = \frac{1}{I_{abs}} + \frac{k_q/k_f}{I_{abs}} [Q].$$

A plot of $1/I_f$ versus [Q] is linear with slope and intercept as $(k_q/k_f)/I_{abs}$ and $1/I_{abs}$, respectively.

Stokes Equation

The Stokes equation is

$$u = \frac{F}{6\ \pi\eta r}$$

where u is the uniform velocity with which a steel ball of radius r falls through a liquid of viscosity η and F is force acting downward on the ball and is given as

$$F=\frac{4}{3}\pi r^3 g\,(\rho_s-\rho_l)$$

where ρ_s and ρ_l are the densities of steel ball and liquids, respectively.

Stokes Lines

Raman lines having lower energies than the energy of incident radiation are known as Stokes lines. These lines correspond to the indirect excitation of the molecule from lower level to high vibration or/and rotational levels.

Surface Active Substances

Substances which produce a marked reduction in surface tension are known to be surface-active substances or surfactants. The reduction in surface tension is due to the fact that the substance has more concentration at the surface as compared to the bulk of solution. The common examples include soaps and detergents.

Surface-Inactive Substances

Surface-inactive substances produces an increase in the surface tension of the solution. This is due to the fact that the substance has more concentration in the bulk as compared to that at the surface. The common examples are inorganic salts.

Surface Tension

The surface tension of a liquid is defined as the force acting along the surface of a liquid at right angle to any line of unit length. The units of surface tension are dyn cm^{-1} (in CGS units) or N m^{-1} (in SI units). Obviously, 1 dyn $cm^{-1}=10^{-3}$ N m^{-1}.

Symmetrical Stretching Vibration of Simple Molecules

Stretching vibration involves the change in the internuclear distance. Symmetrical stretching vibration is observed when the outer atoms move in opposite directions and the central one is stationary.

T

Temperature Dependence of Standard Equlibrium Constant K_p°

The expression is

$$\frac{d \ln K_p^\circ}{dT} = \frac{\Delta \tilde{H}^\circ}{RT^2}$$

The above equation can be written as

$$\frac{d \ln K_p^\circ}{d(1/T)} = \frac{-\Delta H^\circ}{R}$$

The above equation is known as van't Hoff equation.

Thermodynamic Changes during the Adiabatic Change in Volume of an Ideal Gas

For the adiabatic change in volume of an ideal gas, the changes in the thermodynamic variables are as follows.

$$q=0$$

$$w=\Delta U=n\, C_{V,m}\, (T_2-T_1)$$

$$\Delta H=nC_{p,m}\,(T_2-T_1)$$

The relation between T_1 and T_2 are as follows.

(*i*) *Reversible change*

$$TV^{\gamma-1}=\text{constant}$$

$$T^x p^{-R} = \text{constant}$$

$$pV^\gamma = \text{constant}$$

where $x = C_{p,m}$

and $\gamma = C_{p,m}/C_{V,m}$

(*ii*) *Irreversible change*

$$C_{V,m}(T_2 - T_1) = p_{ext}\left(\frac{RT_1}{p_1} - \frac{RT_2}{p_2}\right)$$

For a special case, where $p_{ext} = p_2$ (the pressure of the gas after expansion), we have

$$T_2 = T_1\left(\frac{C_{p,m} + Rp_2/p_1}{C_{p,m}}\right)$$

Thermodynamic Changes during the Isothermal Change in Volume of an Ideal Gas

For the isothermal change in volume of an ideal gas, the changes in the thermodynamic variables are as follows.

(*i*) *Reversible change*

$$q = -w = nRT \ln \frac{p_1}{p_2}$$

$$\Delta U = \Delta H = 0$$

(*ii*) *Irreversible change*

$$q = -w = p_{ext}(V_2 - V_1)$$

$$\Delta U = \Delta H = 0$$

Thermodynamic Data from Cell Potential

The thermodynamic data of cell reaction can be obtained from cell potential by using the following expressions.

$$\Delta G = -nFE$$

$$\Delta H = -nF\left[E - T\left(\frac{\partial E}{\partial T}\right)\right]_p$$

$$\Delta S = nF\left(\frac{\partial E}{\partial T}\right)_p$$

where n is the number of electrons involved in the half-cell reaction.

Thermodynamic Equation of State

The two thermodynamic equations of state are

$$\left(\frac{\partial U}{\partial V}\right)_T = T\left(\frac{\partial p}{\partial T}\right)_V - p$$

$$\left(\frac{\partial H}{\partial p}\right)_T = V - T\left(\frac{\partial V}{\partial T}\right)_p$$

Thermodynamics of Ideal Solution

In making an ideal solution, the following changes in thermodynamic functions are observed.

$$\triangle H_{mix} = 0$$

$$\triangle V_{mix} = 0$$

$$\triangle S_{mix} = -n_{total}\, R \sum_i x_i \ln x_i$$

$$\triangle G_{mix} = n_{total}\, RT \sum_i x_i \ln x_i$$

Third Law of Thermodynamics

The statement of third law of thermodynamics as given by Max Planck is

The minimum entropy may be assigned a zero value for a pure perfectly crystalline substance.

Base on this law, it is possible to find third-law entropy (or simply as entropy) by the use of expression

$$S_T = \int_{0K}^{T} \frac{C_p}{T}\, dT$$

If the pressure throughout is kept at 1 atm, the S_T becomes the standard entropy $S_T°$.

Thermodynamic Process

A thermodynamic process is the path along which a change of state takes place. The different types of processes encountered in thermodynamics are as follows.

Isothermal process. This occurs under constant temperature condition.

Isobaric process. This occurs under constant pressure conditions.

Isochoric proccss. This occurs under constant volume conditions.

Adiabatic process. This occurs under the condition that heat can neither be added to nor removed from the system.

Cyclic process. It is a process in which a sytem undergoes a series of changes and ultimately comes back to the initial state.

Quasi-static (or reversible) process. If a process is carried out in such a way that at any moment the system departs only infinitesimally from an equilibrium state, that process is called quasi-static process. At any instant, the system remains virtually in a state of equilibrium.

Thermodynamic System

Any region of space under investigation is known as system. It can be classified into three categories given below.

Closed System: In this system, matter can neither be added nor removed from it.

Open System: In this system, matter can be added or removed.

Isolated System: This type of system has no interaction with its surroundings. Neither energy not matter can be transferred to or from it.

Tie Line

A line connecting different phases in equilibrium is known as tie line.

Transference Number

The fraction of electricity carried by a particular type of ions in solution is known as its transference number, i.e.

$$t_+ = \frac{I_+}{I} \text{ and } t_- = \frac{I_-}{I}$$

where t_+ and t_- are the respective transference number of cations and anions in a solution containing a known electrolyte, I_+ and I_- are, the charges carried by cations and anions, respectively, and I is the total charge passed into the solution. Obviously,

$$t_+ + t_- = 1$$

Triple Point

At the triple point, three phases, viz. solid, liquid and vapour, of a pure substance are in equilibrium with one another. It is an invariant point. For example, for water such a situation exists at 0.0075 °C and 4.6 mmHg.

Trouton's Rule

The molar entropy of vaporization of most liquids which do not involve hydrogen bonding and also do not possess boiling point less than 150 K is about 10.5 R. Benzene is one of the examples for which we have

$$\triangle H_{vap}(\text{benzene}) = 31.171 \text{ kJ mol}^{-1}$$

$$T_b = 353\text{K}$$

$$\triangle S_{vap}(\text{benzene}) = \frac{31171 \text{ J mol}^{-1}}{353 \text{ K}}$$

$$= 88.3 \text{ J K}^{-1} \text{ mol}^{-1}$$

Tyndall Effect

When a strong converging beam of light is passed through a colloidal solution placed in a dark place, the path of the beam gets illuminated by a bluish light. This phenomenon is known as Tyndall effect. This effect is due to the scattering of light by the colloidal particles.

U

Uncertainty Principle

Based on theoretical reasoning, Heisenberg postulated the following fact regarding the behaviour of subatomic particles. It is not possible to design an experiment with the help of which one can determine simultaneously the precise values of both the position and the momentum of subatomic particles.

Mathematically, uncertainty principle is stated as

$$\triangle p \, \triangle x \leqslant (h/4\pi)$$

where $\triangle p$ and $\triangle x$ are the uncertainties in momentum and position, respectively.

Unit Cell

It is possible to construct a parallelopipeds by connecting lattice points. Each of the parallelopipeds contains a complete unit of pattern of the crystal. By translation or stacking of the parallelopipeds the entire crystal structure can be generated. Such a parallelopiped can be drawn from any crystal lattice, and is called a unit cell.

Valence Bond Method

It is one of the methods to explain the bonding in molecules. This method is very similar to classical chemical concept of valence bond between two atoms and to the electron-pair bond postulated by Lewis, In this method, the molecule is considered to be a collection of atoms and then the interactions between different atoms are considered. As the two atoms of a diatomic molecule is brought closer to each other, interaction amongst electrons and nucleus of atom with other sets it. On the average attractions between electrons and nuclei are larger than repulsions between electrons and electrons, and nucleus and nucleus. Consequently, the energy is lowered. This lowering of energy is continued right up to the stable inter-nuclear distance. Beyond this, repulsive forces are more strong than attractive forces. Hence, energy starts rising. At the internuclear distance, the energy of the molecule is lowest. Hence, it represents the stablest configuration of the molecule.

Valence Shell Electron Repulsion Theory

A simple theory to account for the molecular shape of covalent molecules was developed by Gillespie and Nyholm. This theory, known as VSEPR theory, predicts the shape of molecule by considering the most stable configuration of the bond angles in the molecule. The guiding rules of this theory are as follows:

1. Electron pairs in the valence shell of the central atom of a molecule, whether bonding or lone pair, are regarded as occuping localized orbitals. These orbitals arrange themselves in space so as to minimize the mutual electronic repulsions.
2. The magnitude of the different types of electronic repulsions follow the order given below.

lone pair-lone pair > lone pair-bonding pair > bonding pair-bonding pair

Van der Waals Equation of State

This is an expression which is applicable to a real gas (or a van der Waals gas). Its form is

$$\left(p+\frac{n^2a}{V^2}\right)(V-nb)=nRT$$

where p is the pressure of the gas,

V is the volume of the gas,

T is the temperature of the gas,

n is the amount of the gas,

and a and b are the constants characteristic of the gas.

Van der Waals Interactions

The forces which keep the neutral molecules together in the liquid or solid phases are collectively known as van der waals forces. These forces may be classified into three categories.

Dipole-dipole forces: This exists between the molecules having permanent dipole moments.

Dipole-induced dipole forces: A molecule with a permanent dipole moment can induce a dipole in another molecule and this is followed by attraction between the two molecules.

Dispersion forces: The forces that exist between inert gases are called dispersion forces. The forces of attraction in such molecules (and also in other types of molecules) are explained on the basis that the motion of an electronic cloud at any instant may be slightly displaced relative to the positive nucleus. The atom thus acquires an instantaneous dipole moment which can induce another atom nearby and hence, an instantaneous net attraction exists between the two atoms.

Van't Hoff Equation

See temperature dependence of standard equilibrium constant $K_p°$.

Van't Hoff Factor

The colligative properties of electrolytic solution are found to be larger than the nonelectrolytic solution of corresponding composition. These properties are expressed by the van't Hoff factor defined as follows.

$$i = \frac{\text{colligative effect produced by a given concentration of an electrolytic solution}}{\text{colligative effect produced by the same concentration of a nonelectrolytic solution}}$$

Thus, we may write

$$i=\frac{(-\triangle T_f)}{(-\triangle T_f)_o}=\frac{\triangle T_b}{(\triangle T_b)_o}=\frac{\Pi}{(\Pi)_o}$$

where the quantities without subscript refer to electrolytic solution those with subscript refer to nonelectrolytic solution of the same concentration.

Vapour Pressure

The vapour pressure of a liquid may be defined as the pressure of the vapour in equilibrium with the liquid.

Variation of Entropy with Temperature and Pressure

The expression for the variation of entropy with temperature and pressure is,

$$\mathrm{d}S=\frac{n\,C_{p,m}}{T}\,\mathrm{d}T-\left(\frac{\partial V}{\partial T}\right)_p\mathrm{d}p$$

For an ideal gas

$$\left(\frac{\partial V}{\partial T}\right)_p=-\frac{nR}{p}$$

Hence,

$$\mathrm{d}S=\frac{nC_{p,m}}{T}\,\mathrm{d}T+\frac{nR}{p}\,\mathrm{d}p$$

$$\triangle S=n\left[C_{p,m}\ln\frac{T_2}{T_1}-R\ln\frac{p_2}{p_1}\right]$$

Variation of Entropy with Temperature and Volume

The expression for the variation of entropy with temperature and volume is,

$$\mathrm{d}S=\frac{nC_{V,m}}{T}\,\mathrm{d}T+\left(\frac{\partial p}{\partial T}\right)_V\mathrm{d}V$$

For an ideal gas

$$\left(\frac{\partial p}{\partial T}\right)_V=\frac{nR}{V}$$

Hence,

$$\mathrm{d}S=\frac{nC_{V,m}}{T}\,\mathrm{d}T+\frac{nR}{V}\,\mathrm{d}V$$

$$\triangle S=n\left[C_{V,m}\ \ln\ \frac{T_2}{T_1}R+\ln\frac{V_2}{V_1}\right]$$

Variation of Gibbs Function at Constant Temperature

The required expression is

$$\mathrm{d}G_T=V\mathrm{d}p$$

For a condensed phase

$$G_2=G_1+V(p_2-p_1)$$

and for an ideal gas

$$G_2=G_1+nRT\ \ln\frac{p_2}{p_1}$$

or $$G_2=G_1+nRT\ln\frac{V_1}{V_2}$$

Vibration—Rotation Spectrum

A spectrum exhibiting simultaneous changes in vibrational and rotational levels is known as vibration-rotation spectrum.

Vibrational Spectrum of a Molecule

The vibrational spectrum of a molecule includes absorptions due to the promotion of the molecule from lower vibration level to the higher vibration level. For diatomic molecule, the energy expression of vibrational levels is

$$E=(v+\tfrac{1}{2})\ h\nu[1-(v+\tfrac{1}{2})x_e]$$

where v is the vibrational quantum number, ν is the classical frequency of vibration $(=(^1/_2\pi)\sqrt{k/\mu})$ and x_e is the anharmonicity constant. The energy difference in the vibrational levels lies in the infrared region of the radiation. Hence, absorption of the appropriate frequency in this region produces infrared spectrum of the molecule. The number of vibrational modes in a molecule is,

$3N-5$ for linear molecules

$3N-6$ for non-linear molecules

where N is the number of atoms in the molecule. Each vibration modes produces separate characterestic absorptions. The different groups present in a molecule exhibits characteristic absorptions which practically remains unaffected from one molecule to another. Hence, infrared absorption spectrum is used to identify the tpye of groups present in a molecule.

Virial Equation of State

This is an expression applicable to a real gas. Its form is

$$Z=\frac{pV_m}{RT}=1+\frac{B}{V_m}+\frac{C}{V^2_m}+\frac{D}{V^3_m}+\ldots$$

where B, C,... are temperature dependent constants known as second, third, etc. virial coefficients.

An alternate form of virial equation of state is

$$Z=1+\frac{A_1}{p}+\frac{A_2}{p^2}+\ldots$$

where A_1, A_2, etc. are constants.

Volt

The current is forced through the circuit by an electrical potential difference which is measured in volts (symbol: V). It takes 1 joule of work to move 1 coulomb from a lower to higher potential when the potential difference is 1 V. Thus

$$1 \text{ V}=1 \text{ J}/1 \text{ C}$$

or

$$1 \text{ J}=1 \text{ V C}$$

Walden's Rule

The product of molar conductivity and viscosity of the solution has a constant value, i.e.

$$\Lambda\eta=\text{constant}$$

This rule is known as Walden's rule and is known to be followed by electrolytes containing large ions.

Walsh Rules

A.D. Walsh correlated the molecular orbitals for bent and linear AH_2 molecules and rationalized the following facts regarding the shapes of molecules in the ground state.

AH_2 molecules containing 4 valence electrons should be linear whereas those containing 5-8 valence electrons should be bent.

A similar study on AB_2 and BAC molecules by Walsh rationalized the following facts regarding their shapes.

The molecules with not more than 16 valence electrons are linear in their ground states, the molecules with 17, 18, 19, or 20 valence electrons are bent in their ground states, the apex angle decreases markedly from 16- to 17- and from 17- to 18-electron molecules and less markedly from 18- to 19- and 19- to 20- electron molecules; the molecules with 22 electrons are linear or very linear in their ground states.

For molecules HAB, Walsh obtained the following facts.

The molecules having 10 and 16 valence electrons are linear whereas those having 11-14 electrons are bent.

Wave-Particle Duality

All material particles whether big or small, possess the wave characteristics. This fact was proposed by de Broglie. The relation connecting wave and particle natures is

$$p = \frac{h}{\lambda}$$

where p is the momentum, h is planck's constant and λ is the wavelength of the associated wave.

Weiss Indices

The ratios of the three intercepts h', k' and l' on the three basis directions and the corresponding unit lengths *a*, *b*, and *c* are known as weiss indices.

Wien Effect

In very high electrical fields, $E > 10^5$ V cm^{-1}, an increase in conductivity is observed. This effect is known as Wien effect. This fact has been explained on the basis that at very high fields, the ions move so rapidly that it effectively losses its atmosphere.

X

X-rays

Röntgen observed that when cathode rays are allowed to strike the metallic anode, a new radiations, called X-rays are produced. These are electromagnetic radiations of very short wavelength.

Z

Zeeman Effect

Zeeman in 1896 observed that many of the spectral lines of an atom splits into more than one line when placed in a magnetic field. This effect, known as Zeeman effect, has been explained by the fact that the degeneracy of orbitals having magnetic number *m* greater than zero is lifted up in the presence of a magnetic field.

When the Zeeman effect is examined under high resolution microscope, the spectrum is found to be more complicated. In fact, a spectral line which appears to be a single one, actually consists of more lines closely placed to each other. This fact, known as anomalous Zeeman effect, has been explained on the basis of spin-orbit coupling of angular momentum vectors.

Zero Point Energy

The molecule in the lowest vibrational level have some energy which is known as zero point energy. Its exprcssion

$$E^{\circ}=\tfrac{1}{2}hv\,(1-\tfrac{1}{2}\,x_e)$$

where v is the frequency of vibration $(=(1/2\pi)\sqrt{k/\mu})$ and x_e is the anharmonicity constant. The zero point energy in the molecule is in agreement with uncertainty principle.

Zeroth Law of Thermodynamics

This law states that "two systems in thermal equilibrium with a third system are also in thermal equilibrium with each other". Recording of temperature of a system by a thermometer is also based on this law. When a thermometer is placed in the system, it comes to thermal equilibrium with the latter and thus records a constant value.

Zinc Blende Structure

In zinc blende structure, anions occupy face-centred cubical positions and cations occupy half of the tetrahedral holes in it.

Zone Refining

The technique of zero refining is based on the principle of fractional crystallization. It is employed to remove impurities from a given substance.

Appendices

Table A.1: Seven Base Quantities in SI Units

Physical quantity	*Name of unit*	*Symbol of unit*	*Definition*
Length	meter	m	1 650 763.73 wavelengths in vacuum of the radiation corresponding to the transition $2p_{10}$—$5d_5$ of the krypton-86
Mass	kilogram	kg	A cylinder of platinum-irridium alloy kept by the International Bureau of Weights and Measures in Paris
Time	second	s	The duration of 9 192 631 770 cycles of the radiation associated with the transition between the two hyperfine levels of the ground caesium-133 atom
Electric current	ampere	A	The magnitude of the current that, when flowing through each of two long parallel wires separated by one 1 m in the free space, results in a force between the two wires of 2×10^{-7} N for each meter of length
Thermodynamic temperature	kelvin	K	Origin is at absolute zero and the triple point of water is 273.16 K
Amount of substance	mole	mol	Amount of substance that contains as many elementary entities (atoms, molecules, ions, etc.) as there are carbon atoms in 0.012 kg of carbon-12
Luminous intensity	candela	cd	The luminous intensity in the perpendicular direction, of a surface of 1/600 000 sq m of a black body at the temperature of freezing platinum under a pressure of 101.325 kPa.

Table A.2: Units Derived from the Base SI Units

Physical quantity	*Name of unit*	*Symbol and definition*
Force	Newton	N=kg m s^{-2} or J m^{-1}
Energy	Joule	J=kg m^2 s^{-2} or N m
Electric charge	Coulomb	C=A s
Potential difference	Volt	V=kg m^2 s^{-3} A^{-1} or J A^{-1} s^{-1}
Resistance	Ohm	Ω=kg m^2 s^{-3} A^{-2} or V A^{-1}
Frequency	Hertz=cycle per second	Hz=s^{-1}
Area	Square metre	m^2
Volume	Cubic metre	m^3
Density	Kilogram per cubic metre	kg m^{-3}
Velocity	Metre per second	m s^{-1}
Angular velocity	Radian per second	rad s^{-1}
Acceleration	Metre per second	m s^{-2}
Pressure	Newton per square metre or pascal	N m^{-2} or Pa
Conductivity	Siemen	S=Ω^{-1}
Magnetic flux density	Tesla	T=Wb m^{-2}=V s m^{-2}
Electric capacitance	Farad	F=C V^{-1}
Magnetic flux	Weber	Wb=V s
Inductance	Henry	H=Wb A^{-1}

Table A.3: CGS Units vis-a-vis SI Units

Physical quantity	*CGS units*		*SI units*	
	Name	*Symbol*	*Name*	*Symbol*
Length	centimetre	cm	metre	m
	Angstron (10^{-8} cm)	Å		
Mass	gram	g	kilogram	kg
Time	second	sec	second	s
Temperature	celsius	°C		
	kelvin	°K	kelvin	K
Energy	calorie	cal	joule	J
	kilocalorie	kcal	kilojoule	kJ
	litre-atmosphere	lit-atm		
	ergs	erg		
Electric current	ampere	A	ampere	A

Table A.4: Conversion of CGS Units to SI Units

Quantity	*Units*	*Equivalent**
Length	Angtorn, Å	10^{-10} m$=10^{-1}$ nm$=10^{2}$pm
	micron, μ	10^{-6} m
Volume	litre	10^{-3} m^{3}=dm^{3}
Force	dyne	10^{-5} N
Energy	erg	10^{-7} J
	cal	4.184 J
	eV	1.602 1×10^{-19} J
	eV/mol	98.484 kJ myl^{-1}
Pressure	atmosphere	101.325 kN m^{-2}
	mmHg (or Torr)	133.322 N m^{-2}
	bar (10^{6} dynes/cm^{2})	10^{5} N m^{-2}
Viscosity	poise	10^{-1} kg m^{-1} s^{-1}
Magnetic flux density (magnetic induction)	gauss	10^{-4} T

*Symbols used for fractions and multiples are given in the next Table.

Table A.5: SI Prefixes

Fraction	*Prefix*	*Symbol*	*Multiples*	*Prefix*	*Symbol*
10^{-1}	deci	d	10	deca	da
10^{-2}	centi	c	10^{2}	hecto	h
10^{-3}	milli	m	10^{3}	kilo	k
10^{-6}	micro	μ	10^{6}	mega	M
10^{-9}	neno	n	10^{9}	giga	G
10^{-12}	pico	p	10^{12}	tera	T
10^{-15}	femto	f	10^{15}	peta	P

Table A:6: Values of Some Physico-Chemical Constants

Constant	*CGS units*	*SI units*
Acceleration due to gravity, g	980.665 cm s^{-2}	9.806 65 m s^{-2}
Atomic mass unit, m_{au}	1.660 56 $\times 10^{-24}$ g	1.660 56 $\times 10^{-27}$ kg
Avogadro constant, N_A	6.02205 $\times 10^{23}$ molecules mol^{-1}	6.022 05 $\times 10^{23}$ mol^{-1}
Bohr magneton, μ_B	9.274 1 $\times 10^{-21}$ erg $gauss^{-1}$	9.274 09 $\times 10^{-24}$ J T^{-1}
Bohr radius, a_0	0.529 177 Å	5.291 77 $\times 10^{-11}$ m
Boltzmann constant, k	1.380 66 $\times 10^{-16}$ erg K^{-1}	1.380 66 $\times 10^{-23}$ J K^{-1}
Electronic charge, e	4.802 98 $\times 10^{-10}$ esu	1.602 16 $\times 10^{-19}$ C
Electronic rest mass, m_e	9.109 53 $\times 10^{-28}$ g	9.109 53 $\times 10^{-31}$ kg
Faraday constant, F	96 487 coulomb $equiv^{-1}$	9.648 46 $\times 10^{4}$ C mol^{-1}
Gas constant, R	8.314 41 $\times 10^{7}$ ergs K^{-1} mol^{-1}	8.314 41 J K^{-1} mol^{-1} 8.314 41 Pa m^3 K^{-1} mol^{-1}
	0.082 054 litre-atm K^{-1} mol^{-1}	8.314 41 kPa dm^3 K^{-1} mol^{-1}
	1.987 cal K^{-1} mol^{-1}	8.31441 MPa cm^3 K^{-1} mol^{-1}
		0.08314 bar dm^3 K^{-1} mol^{-1}
Molar volume of ideal gas at 0°C and 1 atm, V_m	22.414 litres	2.241 4 $\times 10^{-2}$ m^3 mol^{-1}
Planck's constant, h	6.626 18 $\times 10^{-27}$ erg s	6.626 18 $\times 10^{-34}$ J s
Proton rest mass, m_p	1.672 65 $\times 10^{-24}$ g	1.672 65 $\times 10^{-27}$ kg
Vacuum speed of light, c	2.997 925 $\times 10^{10}$ cm s^{-1}	2.997 925 $\times 10^{8}$ m s^{-1}
Standard atmospheric pressure	76 cmHg 760 mmHg (or Torr) 1.0132 $\times 10^{6}$ dyn/cm^2	101.325 kPa 1.013 25 bar

Table A.7: Relative Atomic Masses of Elements

Element	*Symbol*	*Atomic Number*	*Relative Atomic Mass**
Actinium	Ac	89	(227)
Aluminium	Al	13	27.0
Americium	Am	95	(243)
Antimony	Sb	51	121.8
Argon	Ar	18	39.9
Arsenic	As	33	74.9
Astatine	At	85	(210)
Barium	Ba	56	137.3
Berkelium	Bk	97	(245)
Beryllium	Be	4	9.01
Bismuth	Bi	83	209.0
Boron	B	5	10.8
Bromine	Br	35	79.9
Cadmium	Cd	48	112.4
Caesium	Cs	55	132.9
Calcium	Ca	20	40.1
Californium	Cf	98	(251)
Carbon	C	6	12.0
Cerium	Ce	58	140.1
Chlorine	Cl	17	35.5
Chromium	Cr	24	52.0
Cobalt	Co	27	58.9
Copper	Cu	29	63.5
Curium	Cm	96	(245)
Dysprosium	Dy	66	162.5
Einsteinium	Es	99	(254)

(*Contd.*)

Element	*Symbol*	*Atomic Number*	*Relative Atomic Mass**
Erbium	Er	68	167.3
Europium	Eu	63	152.0
Fermium	Fm	100	(254)
Fluorine	F	9	19.0
Francium	Fr	87	(223)
Gadolinium	Gd	64	157.3
Gallium	Ga	31	69.7
Germanium	Ge	32	72.6
Gold	Au	79	197.0
Hafnium	Hf	72	178.5
Helium	He	2	4.00
Holmium	Ho	67	164.9
Hydrogen	H	1	1.008
Indium	In	49	114.8
Iodine	I	53	126.9
Iridium	Ir	77	192.2
Iron	Fe	26	55.8
Krypton	Kr	38	83.8
Lanthanum	La	57	138.9
Lawrencium	Lr	103	(257)
Lead	Pb	82	207.2
Lithium	Li	3	6.94
Lutetium	Lu	71	175.0
Magnesium	Mg	12	24.3
Manganese	Mn	25	54.9
Mendelevium	Md	101	(256)
Mercury	Hg	80	200.6

(Contd.)

Element	Symbol	Atomic Number	Relative Atomic Mass*
Molybdenum	Mo	42	95.9
Neodymium	Nd	60	144.2
Neon	Ne	10	20.2
Neptunium	Np	93	237.0
Nickel	Ni	28	58.7
Niobium	Nb	41	92.9
Nitrogen	N	7	14.0
Nobelium	No	102	(254)
Osmium	Os	76	190.2
Oxygen	O	8	16.0
Palladium	Pd	46	106.4
Phosphorus	P	15	31.0
Platinum	Pt	78	195.1
Plutonium	Pu	94	(242)
Polonium	Po	84	(210)
Potassium	K	19	39.1
Praseodymium	Pr	59	140 9
Promethium	Pm	61	(145)
Protactinium	Pa	91	231.0
Radium	Ra	88	226.0
Radon	Rn	86	(222)
Rhenium	Re	75	186.2
Rhodium	Rh	45	102.9
Rubidium	Rb	37	85.5
Ruthenium	Ru	44	101.1
Samarium	Sm	62	150.4

(*Contd.*)

Element	*Symbol*	*Atomic Number*	*Relative Atomic Mass**
Scandium	Sc	21	45.0
Selenium	Se	34	79.0
Silicon	Si	14	28.1
Silver	Ag	47	107.9
Sodium	Na	11	23.0
Strontium	Sr	38	87.6
Sulphur	S	16	32.1
Tantalum	Ta	73	180 9
Technetium	Tc	43	98.9
Tellurium	Te	52	127.6
Terbium	Tb	65	158.9
Thallium	Ti	81	204.4
Thorium	Th	90	232.0
Thulium	Tm	69	168.9
Tin	Sn	50	118.7
Titanium	Ti	22	47.9
Tungsten	W	74	183.8
Uranium	U	92	238.0
Vanadium	V	23	50.9
Xenon	Xe	54	131.3
Ytterbium	Yb	70	173.0
Yttrium	Y	39	88.9
Zinc	Zn	30	65.4
Zirconium	Zr	40	91.2

*Number in parentheses give the mass number of the most stable isotope.

Table A.8: Critical Constants

Gas	Critical Constants		
	p_c / kPa	V_c / $dm^3\ mol^{-1}$	T_c / K
He	228.99	0.0578	5.3
H_2	1296.96	0.0650	33.3
Ne	2634.32	0.0417	44.5
A	4863.60	0.0752	151
Xe	5865.70	0.1202	289.81
N_2	3394.39	0.0901	126.1
O_2	5035.85	0.0744	154.4
CH_4	4640.69	0.0990	190.7
CO_2	7376.46	0.4942	304.2
C_2H_6	4883.87	0.139	305.5
C_3H_8	4265.78	0.195	370.0
n-Butane	3647.7	0.250	426
Ethylene	5116.92	0.126	282.8
Acetylene	6241.62	0.113	308.6
H_2O	22058.45	0.0566	647.3
NH_3	11368.67	0.0720	405.5
CH_3OH	7971.24	0.118	513.2

Table A.9: Van der Waals Constants

Gas	a kPa dm^3 mol^{-2}	b dm^3 mol^{-1}
H_2	12.764	0.026 61
He	3.457	0.023 70
N_2	140.842	0.039 13
O_2	137.802	0.031 83
Cl_2	657.903	0.056 22
NO	135.776	0.027 89
NO_2	535.401	0.044 24
H_2O	553.639	0.030 49
CH_4	228.285	0.042 78
C_2H_6	556.173	0.063 80
C_3H_8	877.880	0.084 45
$C_4H_{10}(n)$	1 466.173	0.122 6
C_4H_{10}(iso)	1 304.053	0.114 2
$C_5H_{15}(n)$	1 926.188	0.146 0
CO	150.468	0.039 85
CO_2	363.959	0.042 67

Table A.10: Coefficients of Viscosity of Some Common Liquids At 20 °C

Substance	*Coefficient of viscosity* (η/poise)$\times 10^3$ *or* (η/N m^{-2} s)$\times 10^4$	*Substance*	*Coefficient of viscosits* (η/poise)$\times 10^3$ *or* (η/N m^{-2} s)$\times 10^4$
Acetic acid	12.29	Ethyl ether	2.33
n-Butyl alcohol	29.5	Acetone	3 29
Ethyl alcohol	12.0	Carbon tetra-chloride	9.68
Methyl alcohol	5.92	Chloroform	5.63
Water	10.02	Benzene	6.47
Nitrobenzene	20.10	Toluene	5.90
Ethylene glycol	199		
Glycerol	8500		

Table A.11: Surface Tensions of Some Common Liquids At 20 °C

Substance	*Surface tension* γ/dyn cm^{-1} *or* $\gamma\times 10^3$/N m^{-1}	*Substance*	*Surface tension* γ/dyn cm^{-1} *or* $\gamma\times 10^3$/N m^{-1}
Water	72.8	Acetic acid	27.42
Benzene	28.87	Methyl alcohol	22.55
Toluene	28.53	Ethyl alcohol	22.30
Carbon disulphide	32.25	Ethyl ether	17.05
Chloroform	27.2	Nitrobenzene	43.35
Carbon tetra-chloride	26.75	*n*-Butyl alcohol	24.52
		n-Propyl alcohol	23.75
Acetone	23.32	*o*-Xylene	30.03
Ethyl acetate	23.75	*p*-Xylene	28.31
Methyl acetate	24.8	Chlorobenzene	33.25

Table A.12: Ionization Constants of Common Acids At 25 °C

Acid	*Equilibria*	K_a/mol dm^{-3}
Nitric	$HNO_3 \rightarrow H^+ + NO_3^-$	$\geqslant 10^2$
Hydrochloric	$HCl \rightarrow H^+ + Cl^-$	$\geqslant 10^7$
Hydrobromic	$HBr \rightarrow H^+ + Br^-$	$\geqslant 10^9$
Hydroiodic	$HI \rightarrow H^+ + I^-$	$\geqslant 10^{11}$
Sulphuric	$H_2SO_4 \rightarrow H^+ + HSO_4^-$	$\geqslant 10^{10}$
Acetic	$HAc \rightleftharpoons H^+ + Ac^-$	1.8×10^{-5}
Benzoic	$C_7H_5O_2H \rightleftharpoons H^+ + C_7H_5O_2^-$	6.0×10^{-5}
Chlorous	$HClO_2 \rightleftharpoons H^+ + ClO_2^-$	1.1×10^{-2}
Formic	$HCO_2H \rightleftharpoons H^+ + HCO_2^-$	1.8×10^{-4}
Hydrazoic	$HN_3 \rightleftharpoons H^+ + N_3^-$	1.9×10^{-5}
Hydrocyanic	$HCN \rightleftharpoons H^+ + CN^-$	4.0×10^{-10}
Hydrofluoric	$HF \rightleftharpoons H^+ + F^-$	6.7×10^{-4}
Hypobromous	$HOBr \rightleftharpoons H^+ + OBr^-$	2.1×10^{-9}
Hypochlorous	$HOCl \rightleftharpoons H^+ + OCl^-$	3.2×10^{-8}
Nitrous	$HNO_2 \rightleftharpoons H^+ + NO_2^-$	4.5×10^{-4}
Arsenic	$H_3AsO_4 \rightleftharpoons H^+ + H_2AsO_4^-$	2.5×10^{-4}
	$H_2AsO_4^- \rightleftharpoons H^+ + HAsO_4^{2-}$	5.6×10^{-8}
	$HAsO_4^{2-} \rightleftharpoons H^+ + AsO_4^{3-}$	3×10^{-13}
Carbonic	$CO_2 + H_2O \rightleftharpoons H^+ + HCO_3^-$	4.2×10^{-7}
	$HCO_3^- \rightleftharpoons H^+ + CO_3^{2-}$	4.8×10^{-11}
Hydrosulphuric	$H_2S \rightleftharpoons H^+ + HS^-$	1.1×10^{-7}
	$HS^- \rightleftharpoons H^+ + S^{2-}$	1.0×10^{-14}
Oxalic	$H_2C_2O_4 \rightleftharpoons H^+ + HC_2O_4^-$	5.9×10^{-2}
	$HC_2O_4^- \rightleftharpoons H^+ + C_2O_4^{2-}$	6.4×10^{-5}

(*Contd.*)

Acid	*Equilibria*	K_a/mol dm^{-3}
Phosphoric	$H_3PO_4 \rightleftharpoons H^+ + H_2PO_4^-$	7.5×10^{-5}
	$H_2PO_4^- \rightleftharpoons H^+ + HPO_4^{2-}$	6.2×10^{-8}
	$HPO_4^{2-} \rightleftharpoons H^+ + PO_4^{3-}$	1×10^{-12}
Phosphorous	$H_3PO_3 \rightleftharpoons H^+ + H_2PO_3^-$	1.6×10^{-2}
	$H_2PO_3^- \rightleftharpoons H^+ + HPO_3^{2-}$	7×10^{-7}
Sulphuric	$H_2SO_4 \rightarrow H^+ + HSO_4^-$	strong
	$HSO_4^- \rightleftharpoons H^+ + SO_4^{2-}$	1.3×10^{-2}
Sulphurous	$SO_2 + H_2O \rightleftharpoons H^+ + HSO_3^-$	1.3×10^{-2}
	$HSO_3^- \rightleftharpoons H^+ + SO_3^{2-}$	5.6×10^{-8}

Table A.13: Ionization Constants of Common Bases At 25 °C

Base	*Equilibria*	K_b/mol dm^{-3}
Ammonia	$NH_3 + H_2O \rightleftharpoons NH_4^+ + OH^-$	1.8×10^{-5}
Aniline	$C_6H_5NH_2 + H_2O \rightleftharpoons C_6H_5NH_3^+ + OH^-$	4.6×10^{-10}
Dimethylamine	$(CH_3)_2NH + H_2O \rightleftharpoons (CH_3)_2NH_2^+ + OH^-$	7.4×10^{-4}
Hydrazine	$N_2H_4 + H_2O \rightleftharpoons N_2H_5^+ + OH^-$	9.8×10^{-7}
Methylamine	$CH_3NH_2 + H_2O \rightleftharpoons CH_3NH_3^+ + OH^-$	5.0×10^{-4}
Pyridine	$C_5H_5N + H_2O \rightleftharpoons C_5H_5NH^+ + OH^-$	1.5×10^{-9}
Trimethylamine	$(CH_3)_3N + H_2O \rightleftharpoons (CH_3)_3NH^+ + OH^-$	7.4×10^{-5}

Table A.14: Some of the Recommended Buffers and their pH Ranges

Constituents	pH-*range*
Glycine and glycine hydrochloride	1.0-3.7
Phthalic acid and potassium acid phthalate	2.2-3.8
Acetic acid and sodium acetate	3.7-5.6
Disodium citrate and trisodium citrate	5.0-6.3
Monosodium phosphate and disodium phosphate	5.8-8.0
Boric acid and borax	68.-9.2
Borax and sodium hydroxide	9.2-11.0
Disodium phosphate and trisodium phosphate	11.0-12.0

Table A.15: Acid-Base Indicators

Indicator	*Colour change*		pH *transition range*	
	Acid form	*Base form*	*Acid form predominate at* pH	*Base form Predominate at* pH
Picric acid	Colourless	Yellow	0.0	1.2
Malachite green	Yellow	Green	0.0	2.0
Methyl violet	Yellow	Violet	0.1	3.2
m-Cresol purple	Red	Yellow	1.2	2.8
Thymol blue	Red	Yellow	1.2	2.8
Bromophenol blue	Yellow	Blue	3.0	4.6
Congo red	Blue	Red	3.0	5.0
Methyl orange	Red	Yellow	3.1	4.4
Bromocresol green	Yellow	Blue	3.8	5.4
Methyl red	Red	Yellow	4.2	6.3
Litmus	Red	Blue	4.5	8.3
Propyl red	Red	Yellow	4.6	6.4
Chlorophenol red	Yellow	Red	4.8	6.4
Hematoxylin	Yellow	Red	5.0	6.0
p-Nitrophenol	Colourless	Yellow	5.0	7.0
Bromocresol purple	Yellow	Purple	5.2	6.8
Bromothymol blue	Yellow	Blue	6.0	7.6
Phenol red	Yellow	Red	6.8	8.4
m-Cresol purple	Yellow	Purple	7.4	9.0
Thymol blue	Yellow	Blue	8.0	9.6
Phenolphthalein	Colourless	Red	8.3	10.0
Thymolphthalein	Colourless	Blue	9.3	10.5
Alizarin yellow R	Yellow	Lavender	10.0	12.1
Alizarin blue S	Green	Blue	11.0	13.0
Malachite green	Green	Colourless	11.4	13.0
Trinitrobenzene	Colourless	Orange	12.	14.0

Table A.16: Solubility Products At 25°C

Salt	K°_{sp}
$PbBr_2$	4.6×10^{-6}
Hg_2Br_2	1.3×10^{-22}
$AgBr$	5.0×10^{-13}
$BaCO_3$	1.6×10^{-9}
$CdCO_3$	5.2×10^{-12}
$CaCO_3$	4.7×10^{-9}
$CuCO_3$	2.5×10^{-10}
$FeCO_3$	2.1×10^{-11}
$PbCO_3$	1.5×10^{-15}
$MgCO_3$	1.0×10^{-15}
$MnCO_3$	8.8×10^{-11}
Hg_2CO_3	9.0×10^{-17}
$NiCO_3$	1.4×10^{-7}
Ag_2CO_3	8.2×10^{-12}
$SrCO_3$	$7\ \times10^{-10}$
$ZnCO_3$	$2\ \times10^{-10}$
$PbCl_2$	1.6×10^{-5}
Hg_2Cl_2	1.1×10^{-18}
$AgCl$	1.7×10^{-10}
$BaCrO_4$	8.5×10^{-11}
$PbCrO_4$	$2\ \times10^{-16}$
Hg_2CrO_4	$2\ \times10^{-9}$
Ag_2CrO_4	1.9×10^{-12}
$SrCrO_4$	3.6×10^{-5}

(*Contd.*)

Salt	K°_{sp}
HgS	3.0×10^{-53}
CuS	4.0×10^{-38}
PbS	1.0×10^{-29}
SnS	8.0×10^{-29}
Sb_2S_3	10^{-80}
Bi_2S_3	1.6×10^{-72}
CdS	1.0×10^{-28}
ZnS	2.5×10^{-22}
CoS	7×10^{-23}
NiS	3×10^{-21}
MnS	5.6×10^{-16}
FeS	1.0×10^{-19}
BaF_2	2.4×10^{-5}
CaF_2	3.9×10^{-11}
PbF_2	4×10^{-8}
MgF_2	8×10^{-8}
SrF_2	7.9×11^{-11}
$Al(OH)_3$	5×10^{-33}
$Ba(OH)_2$	5×10^{-3}
$Cd(OH)_2$	2×10^{-14}
$Ca(OH)_2$	1.3×10^{-6}
$Cr(OH)_3$	6.7×10^{-31}
$Co(OH)_2$	2.5×10^{-16}
$Co(OH)_3$	2.5×10^{-43}
$Cu(OH)_2$	1.6×10^{-19}
$Fe(OH)_2$	1.8×10^{-15}
$Fe(OH)_3$	6×10^{-38}

Salt	K°_{sp}
$Pb(OH)_2$	4.2×10^{-15}
$Mg(OH)_2$	8.9×10^{-13}
$Mn(OH)_2$	2×10^{-13}
$Ni(OH)_2$	1.6×10^{-16}
$Sr(OH)_2$	3.2×10^{-4}
$Sn(OH)_2$	3×10^{-27}
$Zn(OH)_2$	4.5×10^{-17}
PbI_2	8.3×10^{-9}
Hg_2I_2	4.5×10^{-23}
AgI	8.5×10^{-17}
AgCN	1.6×10^{-14}
AgCNS	1.0×10^{-12}
BaC_2O_4	1.5×10^{-8}
CaC_2O_4	1.3×10^{-9}
PbC_2O_4	8.3×10^{-12}
MgC_2O_4	8.6×10^{-5}
$Ag_2C_2O_4$	1.1×10^{-11}
SrC_2O_4	5.6×10^{-8}
$Ba_3(PO_4)_2$	6×10^{-39}
$Ca_3(PO_4)_2$	1.3×10^{-32}
$Pb_3(PO_4)_2$	1×10^{-54}
$Ag_3(PO_4)$	1.8×10^{-18}
$Sr_3(PO_4)_2$	1×10^{-31}
$BaSO_4$	1.5×10^{-9}
$CaSO_4$	2.4×10^{-5}
$PbSO_4$	1.3×10^{-8}
Ag_2SO_4	1.2×10^{-6}
$SrSO_4$	7.6×10^{-7}

Table A.17: Ionic Molar Conductivities at Infinite Dilution at 25 °C

Cations	$\overset{\infty}{\lambda}_+/\Omega^{-1}$ cm^2 mol^{-1}	*Cations*	$\overset{\infty}{\lambda}_+/\Omega^{-1}$ cm^2 mol^{-1}	*Anions*	$\overset{\infty}{\lambda}_-/\Omega^{-1}$ cm^2 mol^{-1}	*Anions*	$\overset{\infty}{\lambda}_-/\Omega^{-1}$ cm^2 mol^{-1}
H^+	349.8	$\frac{1}{2}Hg^{2+}$	63.3	OH^-	197.6	$H_2AsO_4^-$	34
Li^+	38.66	$\frac{1}{2}Zn^{2+}$	52.8	F^-	55.4	HS^-	64
Na^+	50.11	$\frac{1}{2}Cd^{2+}$	54	Cl^-	76.34	HSO_3^-	58
K^+	73.52	$\frac{1}{2}Pb^{2+}$	70	Br^-	78.14	HSO_4^-	52
Rb^+	77.8	$\frac{1}{2}Mn^{2+}$	53.5	I^-	76.97	$\frac{1}{2}CO_3^{2-}$	69.3
Cs^+	77.3	$\frac{1}{2}Co^{2+}$	55	CN^-	82	$\frac{1}{2}HPO_4^{2-}$	57
NH_4^+	73.4	$\frac{1}{2}Ni^{2+}$	54	CNS^-	66	$\frac{1}{2}SO_3^{2-}$	72
Ag^+	61.92	$\frac{1}{3}Fe^{3+}$	68	NO_2^-	71.44	$\frac{1}{2}SO_4^{2-}$	80
$\frac{1}{2}Mg^{2+}$	53.06	$\frac{1}{3}Al^{3+}$	63	ClO_2^-	52	$\frac{1}{2}S_2O_3^{2-}$	87.4
$\frac{1}{2}Ca^{2+}$	59.50	$\frac{1}{3}Cr^{3+}$	67	ClO_3^-	64.6	$\frac{1}{2}CrO_4^{2-}$	83
$\frac{1}{2}Sr^{2+}$	59.46	$\frac{1}{3}Co(NH_3)_6^{3+}$	99.2	ClO_4^-	67.4	$\frac{1}{3}PO_4^{3-}$	92.8

$\frac{1}{2}Ba^{2+}$	63.64	$N(CH_3)_4^+$	44.92	BrO_3^-	55.7	$\frac{1}{3}Fe(CN)_6^{3-}$	99.1
$\frac{1}{2}Cu^{2+}$	56.6	$NH(CH_3)_3^+$	42	IO_3^-	40.7	$\frac{1}{4}Fe(CN)_6^{4-}$	111
$\frac{1}{2}Hg_2^{2+}$	68.6			IO_4^-	54.5	Formate$^-$	54.6
				$\frac{1}{2}$Oxalate^{2-}	74.1	Acetate$^-$	40.9
				$\frac{1}{2}$Tartrate^{2-}	59.6	Monochloro acetate$^-$	39.8
				$\frac{1}{3}$Citrate^{3-}	71.5	Dichloro acetate$^-$	38
				MnO_4^-	61	Trichloro acetate$^-$	35
				HCO_3^-	44.5	Benzoate$^-$	32.3
				$H_2PO_4^-$	36	*n*-Propionate$^-$	35.8

Table A.18: Standard Half-Cell Potentials at 25 °C

Electrode reaction	*E°/V*	*Half-cell representation*
$Li^+ + e^- = Li$	—3.045	$Li^+ \mid Li$
$K^+ + e^- = K$	—2.925	$K^+ \mid K$
$Na^+ + e^- = Na$	—2.714	$Na^+ \mid Na$
$Mg^{2+} + 2e^- = Mg$	—2.37	$Mg^{2+} \mid Mg$
$H_2 + 2e^- = 2H^-$	—2.25	$H^- \mid H_2 \mid Pt$
$Al^{3+} + 3e^- = Al$	—1.66	$Al^{3+} \mid Al$
$2H_2O + 2e^- = H_2 + 2OH^-$	—0.828	$OH^- \mid H_2 \mid Pt$
$Zn^{2+} + 2e^- = Zn$	—0.763	$Zn^{2+} \mid Zn$
$Cd(NH_3)_4^{2+} + 2e^- = Cd + 4NH_3$	—0.61	$Cd(NH_3)_4^{2+}, NH_3 \mid Cd$
$2CO_2 + 2H^+ + 2e^- = H_2C_2O_4$	—0.49	$H_2C_2O_4, H^+ \mid CO_2 \mid Pt$
$Fe^{2+} + 2e^- = Fe$	—0.44	$Fe^{2+} \mid Fe$
$Cr^{3+} + e^- = Cr^{2+}$	—0.41	$Cr^{3+}, Cr^{2+} \mid Pt$
$Cd^{2+} + 2e^- = Cd$	—0.40	$Cd^{2+} \mid Cd$
$Ag(CN)_2^- + e^- = Ag + 2CN^-$	—0.31	$Ag(CN)_2^-, CN^- \mid Ag$
$Cu(OH)_2 + 2e^- = Cu + 2OH^-$	—0.224	$OH^-, Cu(OH)_2 \mid Cu$
$AgI + e^- = Ag + I^-$	—0.151	$I^- \mid AgI \mid Ag$
$Sn^{2+} + 2e^- = Sn$	—0.136	$Sn^{2+} \mid Sn$
$Pb^{2+} + 2e^- = Pb$	—0.126	$Pb^{2+} \mid Pb$
$Cu(NH_3)_4^{2+} + 2e^- = Cu + 4NH_3$	—0.12	$Cu(NH_3)_4^{2+}, NH_3 \mid Cu$
$Fe^{3+} + 3e^- = Fe$	—0.036	$Fe^{3+} \mid Fe$
$2H^+ + 2e^- = H_2$	0.000	$H^+ \mid H_2 \mid Pt$
$AgBr + e^- = Ag + Br^-$	0.095	$Br^- \mid AgBr \mid Ag$

(*Contd.*)

Electrode reaction	$E°/V$	*Half-cell representation*
$Cu^{2+}+e^-=Cu^+$	0.153	$Cu^{2+},Cu^+ \mid Pt$
$Sn^{4+}+2e^-=Sn^{2+}$	0.15	$Sn^{4+},Sn^{2+} \mid Pt$
$AgCl+e^-=Ag+Cl^-$	0.222	$Cl^- \mid AgCl \mid Ag$
$Hg_2Cl_2+2e^-=2Hg+2Cl^-$	0.267 6	$Cl^- \mid Hg_2Cl_2 \mid Hg(Pt)$
$Cu^{2+}+2e^-=Cu$	0.337	$Cu^{2+} \mid Cu$
$Ag(NH_3)_2^++e^-=Ag+2NH_3$	0.373	$Ag(NH_3)_2^+,NH_3 \mid Ag$
$O_2(g)+2H_2O+4e^-=4OH^-$	0.401	$OH^- \mid O_2 \mid Pt$
$Cu^++e^-=Cu$	0.521	$Cu^+ \mid Cu$
$\frac{1}{2}I_2+e^-=I^-$	0.535 5	$I_2, I^- \mid Pt$
$Fe^{3+}+e^-=Fe^{2+}$	0.771	$Fe^{3+}, Fe^{2+} \mid Pt$
$\frac{1}{2}Hg_2^{2+}+e^-=Hg$	0.789	$Hg_2^{2+} \mid Hg(Pt)$
$Ag^++e^-=Ag$	0.799 1	$Ag^+ \mid Ag$
$O_2+4H^++4e^-=2H_2O$	1.229	$H^+ \mid O_2 \mid Pt$
$Br_2(l)+2e^-=2Br^-$	1.056 2	$Br^-, Br_2 \mid Pt$
$Cr_2O_7^{2-}+14H^++6e^-=2Cr^{3+}$ $+7H_2O$	1.33	$CrO_7^{2-},Cr^{3+},H^+ \mid Pt$
$\frac{1}{2}Cl_2(g)+e^-=Cl^-$	1.359 5	$Cl^- \mid Cl_2 \mid Pt$
$Ce^{4+}+e^-=Ce^{3+}$ (1 mol dm^{-3} H_2SO_4)	1.44	$Ce^{4+}, Ce^{3+} \mid Pt$
$Au^{3+}+3e^-=Au$	1.455	$Au^{3+} \mid Au$
$MnO_4^-+8H^++5e^-$ $=Mn^{2+}+4H_2O$	1.51	$MnO_4, Mn^{2+},H^+ \mid Pt$
$Co^{3+}+e^-=Co^{2+}$	1.82	$Co^{3+},Co^{2+} \mid Pt$

Table A.19: Molar Heat Capacities of Gases at Constant Volume and Pressure

Gas	*Values at* 298 K		$C_{p,m}/\text{J K}^{-1}\ mol^{-1}=a+b\ (T/\text{K})+c(T/\text{K})^2$ *from* 300 K *to* 1500 K		
	$Cv_{,m}/\text{J K}^{-1}$	$C_{p,m}/\text{J K}^{-1}$	a	$b\times10^3$	$c\times10^6$
He	12.468	20.795	20.836	0	0
H_2	12.543	28.870	29.066	−0.836	2.012
N_2	20.711	29.037	26.983	5.910	−0.338
O_2	21.129	29.497	25.503	13.612	−4.256
CO	20.795	29.193	26.537	7.683	−1.172
HCl	20.962	29.497	28.166	1.809	1.547
Cl_2	25.690	34.518	31.696	10.144	−4.038
Br_2			35.241	4.075	−1.487
HBr		29.121	27.521	3.995	0.662
NO			29.372	−1.548	10.652
H_2O	24.811	33.095	30.072	9.930	0.872

SO_2			2.5719	57.923	—38.087
CO_2	28.953	37.489	26.760	42 649	—14.783
H_2S			26.715	23.866	—5.063
NH_3	27.489	36.108	25.895	32 999	—3.046
CH_4	27.573	35.982	14.318	74.663	—17.426
C_2H_2	$\simeq$35.606	$\simeq$43.932	30.673	52.810	—16.272
C_3H_8			1.732	270.747	—94.483
C_2H_4	$\simeq$35.229	$\simeq$43.555	11.841	119.666	—36.510
C_2H_6	44.560	53.179	5.753	175.109	—578.522
C_3H_6			13.611	188.765	—57.488
C_6H_6	$\simeq$73.387	81.672			

Table A.20: Selected Values of Thermodynamic Properties at 298.15 K and 101.325 kPa

Substance	ΔH_f kJ mol^{-1}	ΔG_f° kJ mol^{-1}	S° J K^{-1} mol^{-1}	C_p J K^{-1} mol^{-1}
Inorganic Compounds				
$NaF(s)$	—569.02	—540.99	58.58	46.02
$NaCl(s)$	—411.00	—384.03	72.38	49.71
$NaClO_4(s)$	—385.68	—256.90		
$NaBr(s)$	—359.95	—347.69		
$NaI(s)$	—288.03	—237.23		
$Na_2SO_4(s)$	—1384.49	—1266.83	149.49	127.61
$Na_2SO_4.10H_2O(s)$	—4324.08	—3643.97	592.87	587.43
$Na_2CO_3(s)$	—1130.94	—1047.67	135.98	110.50
$NaHCO_3(s)$	—947.68	—851.86	102.09	87.61
$KF(s)$	—562.58	—533.13	66.57	49.08
$KF.2H_2O(s)$	—1158.97	—1015.46	150.62	
$KCl(s)$	—435.87	—408.32	82.68	51.51
$KBr(s)$	—392.17	—379.20	96.44	53.64
$KI(s)$	—327.65	—322.29	104.35	55.06
$KNO_3(s)$	—492.71	—393.13	132.92	96.27
$KMnO_4(s)$	—813.37	—713.79	171.71	119.24
$Mg(OH)_2(g)$	—924.66	—833.75	63.14	77.03
$MgCO_3(s)$	—1112.94	—1029.26	65.69	75.52
$CaO(s)$	—635.55	—604.17	39.75	42.80
$Ca(OH)_2(s)$	—986.59	—896.76	76.15	84.52
$CaSO_4(s)$	—1432.69	—1320.30	106.69	99.58
$CaSO_4.\frac{1}{2}H_2O(s)$	—1575.15	—1435.20	130.54	119.66
$CaSO_4.2H_2O(s)$	—2021.12	—1795.73	193.97	186.19

Substance	$\triangle H_1$ kJ mol^{-1}	$\triangle G_1°$ kJ mol^{-1}	$S°$ J K^{-1} mol^{-1}	C_p JK^{-1} mol^{-1}
$MnSO_4$(s)	—1063.74	—955.96	112.13	100.16
Fe_2O_3(s)	—822.16	—740.99	89.96	104.60
ZnO(s)	—350.20	—318.19	43.93	40.25
Fe_3O_4(s)	—1120.89	—1014.20	146.44	
$CuSO_4$(s)	—769.86	—661.9	113.39	100.83
$CuSO_4.H_2O$(s)	—1083.66	—917.13	149.79	130.96
$.3H_2O$(s)	—1683.01	—1399.97	225.10	205.02
$.5H_2O$(s)	—2277.98	—1879.87	305.43	281.16
AgCl(s)	—127.03	—109.72	96.11	50.79
O_3(g)	142.26	163.43	237.65	38.16
H_2O(g)	—241.83	—228.59	188.72	33.58
H_2O(1)	—285.84	—237.19	69.94	75.30
H_2O_2(l)	—187.61	—113.97	92.05	
HCl(g)	—92.31	—95.27	186.68	29.12
HCl(aq)	—167.46	—131.17	55.23	
Br_2(g)	30.71	3.14	242.35	35.98
HBr(g)	—36.23	—53.22	198.48	29.12
HI(g)	25.94	1.297	206.33	29.16
S(monoclinic)	0.297	0.096	32.55	23.64
SO_2 (g)	—296.06	—297.86	248.53	39.79
SO_3(g)	—395.18	—370.37	256.23	50.63
H_2S(g)	—20.15	—33.02	205.64	33.97
H_2SO_4(aq)	—907.51	—741.99	17.15	16.74
NO(g)	90.37	86.69	210.62	29.86
NO_2(g)	33.85	51.84	240.45	37.91
N_2O_4(g)	9.66	98.29	304.30	79.08
NH_3(g)	—46.19	—16.64	192.51	35.66
HNO_3(1)	—173.23	—79.91	155.60	109.87
PCl_3(g)	—306.35	—286.27	311.67	

(*Contd.*)

Substance	$\triangle H$ / kJ mol^{-1}	$\triangle G_f^\circ$ / kJ mol^{-1}	S° / J K^{-1} mol^{-1}	C_p / J K^{-1} mol^{-1}
C(diamond)	1.826	2.87	2.44	6.06
CO(g)	—110.52	−137.27	197.91	29.14
CO_2(g)	—393.51	—394.38	213.64	37.13
Organic Compounds				
Methane(g)	—74.85	−50.79	186.19	35.71
Ethane(g)	—84.67	—32.89	229.49	52.66
Benzene C_6H_6(g)	82.93	129.66	269.20	81.67
C_6H_6(1)	49.03	172.80	124.50	
Ethylene C_2H_4(g)	52.28	68.12	219.45	43.56
Acetylene C_2H_2(g)	226.75	209.2	200.82	43.93
Formic acid HCOOH(1)	—409.20	—346.02	128.95	99.04
Acetic acid CH_3COOH(1)	—487.02	—392.46	159.83	123.43
Oxalic acid $(COOH)_2$	—826.76	−697.89	120.08	108.78
Formaldehyde HCHO(g)	—115.90	—109.62	218.66	35.35
Methanol CH_3OH(g)	—201.25	—161.92	237.65	
CH_3OH(1)	—238.64	—166.31	126.78	81.59
Ethyl alcohol C_2H_5OH(g)	—235.31	—168.62	282.00	
C_2H_5OH(1)	—277.63	—174.77	160.67	111.46
Acetaldehyde CH_3CHO(g)	—166.36	—133.72	265.68	62.76
Carbon Tetrachloride CCl_4(g)	—106.69	−64.22	309.41	83.51
CCl_4(1)	—139.50	—68.74	214.43	131.75

(Contd.)

Substance	$\triangle H$ / kJ mol^{-1}	$\triangle G_f^\circ$ / kJ mol^{-1}	S° / J K^{-1} mol^{-1}	C_p / J K^{-1} mol^{-1}
Ions in Solutions				
H^+	0.0	0.0	0.0	0.0
OH^-	—229.94	—157.30	—10.54	—133.89
F^-	—329.11	—276.48	—9.62	—123.43
Cl^-	−167.46	—131.17	55.23	—125.52
Br^-	—120.92	—102.82	80.71	—128.45
I^-	—55.94	—51.67	109.37	—129.70
ClO_4^-	—131.42	−8.368	182.00	
I_3^-	—51.88	—51.51	173.64	
BrO_3^-	—40.17	45.61	162.76	—79.50
HS^-	—17.66	12.59	61.09	
S^{2-}	41.84	83.68		
SO_4^{2-}	—907.51	—741.99	17.15	16.74
SO_3^{2-}	—624.25	—497.06	43.51	
NO_2^-	—106.27	—34.52	125.10	
NO_3^-	—206.57	—110.58	146.44	
NH_4^+	—132.80	—79.50	112.84	
PO_4^{3-}	—1284.07	—1025.5	—217.57	
HCO_3^-	—691.11	—587.06	94.98	
CO_3^{2-}	—676.26	—528.10	—53.14	
Zn^{2+}	—152.42	—147.21	—106.48	
Cd^{2+}	—72.38	—77.74	—61.09	
Cu^{2+}	64.39	64.98	—98.74	
Ag^+	105.90	77.11	73.93	37.66
Mg^{2+}	—461.96	—456.01	—109.62	

(Contd.)

Substance	ΔH / kJ mol^{-1}	ΔG_1° / kJ mol^{-1}	S° / J K^{-1} mol^{-1}	C_p / J K^{-1} mol^{-1}
Ca^{2+}	−542.96	−553.04	−55.23	
Li^+	−278.46	−293.80	14.23	
Na^+	−239.66	−261.87	60.25	
K^+	−251.21	−282.25	102.51	
Cr^{3+}		−215.48		
$Cr_2O_7^{2-}$	−1460.63	−1257.29	213.8	
CrO_4^{2-}	−894.33	−736.80	38.49	
MnO_4^-		−503.75		
Mn^{2+}	−218.82	−223.43	−83.69	
$Ag(CN)_2^-$	269.87	301.46	205.02	
$Au(CN)_2^-$	244.35	215.48	414.22	
CN^-	151.04	165.69	117.99	
Gaseous Atoms				
H	217.94	203.24	114.61	20.79
F	76.57	59.41	158.65	22.75
Cl	121.39	105.40	165.09	21.84
Br	111.75	82.38	174.91	20.79
I	106.62	70.15	180.68	20.79
N	472.65	455.51	153.20	20.79
C	718.38	672.98	157.99	20.84
O	247.52	230.09	160.65	21.91
S	222.80	182.30	167.72	23.68
Na	108.70	78.12	153.62	20.79
K	90.00	61.17	160.23	20.79

(*Contd.*)

Substance	$\triangle H_1$ / kJ mol^{-1}	$\triangle G_1^\circ$ / kJ mol^{-1}	S° / J K^{-1} mol^{-1}	C_p / J K^{-1} mol^{-1}
Elements in their States of Aggregation				
H_2(g)	0.0	0.0	130.59	28.84
F_2(g)	0.0	0.0	203.34	31.46
Cl_2(g)	0.0	0.0	222.95	33.93
Br_2(l)	0.0	0.0	152.30	
I_2(s)	0.0	0.0	116.73	54.98
N_2	0.0	0.0	191.49	29.12
C(graphite)	0.0	0.0	5.694	8.644
O_2	0.0	0.0	205.03	29.36
S(rhombic)	0.0	0.0	31.88	22.59
Na(s)	0.0	0.0	51.04	28.41
K(s)	0.0	0.0	63.60	29.16
Mg(s)	0.0	0.0	32.51	23.89
Ca(s)	0.0	0.0	41.63	26.28
Sr(s)	0.0	0.0	54.39	25.10
Ba(s)	0.0	0.0	66.94	26.36
Fe(s)	0.0	0.0	27.15	25.23

Table A.21: Approximate Bond Enthalpies at 298.15 K (kJ/mol)

	C	N	O	F	P	S	Cl	Br	I
H–	413.38	390.79	462.75	563.17	319.66	339.32	431.79	366.1	298.74
C—	347.69	291.62	351.46	440.99		259.41	328.44	275.73	240.16
C=	615.05	615.05	728.02			476.98			
C≡	811.70	891.19							
N—	291.62	160.67		269.87			199.58		
N=	615.05	418.4							
N≡	891.19	945.58							
O—	351.46		138.91	184.93			202.92		
O=	728.0		494.07						

Table A.22: Vapor Pressure of Water below 100°C

Temperature (t/°C)	*Vapor pressure* (p/Torr)
0.0	4.579
1.0	4.926
2.0	5.294
3.0	5.685
4.0	6.101
5.0	6.543
6.0	7.013
7.0	7.531
8.0	8.045
9.0	8.609
10.0	9.202
11.0	9.844
12.0	10.518
13.0	11.231
14.0	11.987
15.0	12.788
16.0	13.634
17.0	14.530
18.0	15.477
19.0	16.477
20.0	17.535
21.0	18.650
22.0	19.827
23.0	21.068
24.0	22.377
25.0	23.756
26.0	25.209
27.0	26.739
28.0	28.349
29.0	30.043
30.0	31.824
31.0	33.695
32.0	35.663

(*Contd.*)

Temperature (t/°C)	*Vapor pressure* (p/Torr)
33.0	37.729
34.0	39.898
35.0	42.175
36.0	44.563
37.0	47.067
38.0	49.692
39.0	52.442
40.0	55.324
41.0	58.34
42.0	61.50
43.0	64.80
44.0	68.26
45.0	71.88
46.0	75.65
47.0	79.60
48.0	83.71
49.0	88.02
50.0	92.51
51.0	97.20
52.0	102.09
53.0	107.20
54.0	112.51
55.0	118.04
56.0	123.80
57.0	129.82
58.0	136.08
59.0	142.60
60.0	149.38
61.0	156.43
62.0	163.77
63.0	171.38
64.0	179.31
65.0	187.54
66.0	196.09
67.0	204.96

(*Contd.*)

Temperature (t/°C)	Vapor pressure (p/Torr)
68.0	214.17
69.0	223.73
70.0	233.7
71.0	243.9
72.0	254.6
73.0	265.7
74 0	272.2
75.0	289.1
76.0	301.4
77.0	314.1
78.0	327.3
79.0	341.0
80.0	355.1
81.0	369.7
82.0	384.9
83.0	400.6
84.0	416.8
85.0	433.6
86.0	450.9
87.0	468.7
88.0	487.1
89.0	506.1
90.0	525.76
91.0	546.05
92.0	566.99
83.0	588.60
94.0	610.90
95.0	633.90
96.0	657.62
97.0	682.07
98.0	707.27
99.0	733.24
100.0	760.00

Table A.23: Surface Tension and Viscosity of Water at Different Temperatures

t/°C	γ/dyn cm^{-1}	t/°C	η/centipoise
10	74.22	15	1.1404
15	73.49	20	1.0050
18	73.05	21	0.9810
20	72.75	22	0.9579
25	71.97	23	0.9858
30	71.18	24	0.9142
40	69.56	25	0.8937
50	67.91	26	0.8737
60	66.18	27	0.8545
70	64.4	28	0.8360
80	62.6	29	0.8180
100	58.9	30	0.8007
		31	0.7840
		32	0.7679
		33	0.7523
		34	0.7371
		35	0.7225

Table A.24: Viscosity (in centipose) of Aqueous Solutions at 20°C

Per cent by mass	*Ethyl alcohol*	*Glycerol*	*Sucrose*
10	1.538	1.31	
20	2.183	1.76	1.945
30	2.71	2.50	3.187
40	2.91	3.72	6.167
50	2.87	6.00	15.43
60	2.67	10.8	58.49
70	2.370	22.5	481.6
80	2.008	60.1	
90	1.610	218	
100	1.200	1412	

Table A 25: Surface Tensions (in dyn cm^{-1}) of Aqueous Solutions

Mass per cent	*Surface tension*	*Mass per cent*	*Surface tension*
Acetic acid (30°C)		*Acetone* (25°C)	
1	68.0	5	55.5
2.475	64.4	10	48.9
5.001	60.1	20	41.1
10.01	54.6	25	38.3
30.09	43.6	50	30.4
49.96	38.4	75	26.8
69.91	34.3	95	24.2
100	26.6	100	23.2
Methyl alcohol (30°C)		*Ethyl alcohol* (30°C)	
1.011	68.4	0.979	66.1
2.5	65.3	2.143	61.6
4.997	61.0	4.994	54.2
9.994	54.6	10.39	45.9
25.00	43.0	25.00	34.1
50.00	32.9	50.00	27.5
75.00	27.1	75.06	24.7
100	21.8	100	21.5